Predicting and Monitoring Climate Risks Through Green AI:

Weather Forecasting, Disaster Prediction, and Biodiversity Monitoring

Fanar Shwedeh
https://orcid.org/0009-0005-5471-8418
City University, Ajman, UAE

IGI Global Scientific Publishing
Publishing Tomorrow's Research Today

Vice President of Editorial — Melissa Wagner
Director of Acquisitions — Mikaela Felty
Director of Book Development — Jocelynn Hessler
Production Manager — Mike Brehm
Cover Design — Jose Rosado

Published in the United States of America by
IGI Global Scientific Publishing
701 East Chocolate Avenue
Hershey, PA, 17033, USA
Tel: 717-533-8845 | Fax: 717-533-7115
Website: https://www.igi-global.com E-mail: cust@igi-global.com

Library of Congress Cataloging-in-Publication Data

Names: Shwedeh, Fanar, 1983- editor
Title: Predicting and monitoring climate risks through green AI : weather forecasting, disaster prediction, and biodiversity monitoring / Fanar Shwedeh.
Description: Hershey, PA, USA : IGI Global Scientific Publishing, [2026] | Includes bibliographical references. | Summary: "This book explores how AI drives green innovation, from renewable energy and smart cities to agriculture and ecosystem protection. It examines sustainable algorithms for climate prediction, disaster management, and biodiversity monitoring, emphasizing ethical, efficient, and responsible AI as a catalyst for a more resilient and sustainable future"-- Provided by publisher.
Identifiers: LCCN 2025054487 (print) | LCCN 2025054488 (ebook) | ISBN 9798337375540 hardcover | ISBN 9798337375557 paperback | ISBN 9798337375564 ebook
Subjects: LCSH: Climatic changes--Risk assessment--Data processing | Weather forecasting--Data processing | Environmental monitoring--Remote sensing--Data processing | Artificial intelligence--Environmental aspects
Classification: LCC QC903 .P924 2026 (print) | LCC QC903 (ebook)
LC record available at https://lccn.loc.gov/2025054487
LC ebook record available at https://lccn.loc.gov/2025054488
British Cataloguing in Publication Data
A Cataloguing in Publication record for this book is available from the British Library.

Table of Contents

Detailed Table of Contents

Blockchain and artificial intelligence (AI) are increasingly recognized as foundational technologies for advancing environmental sustainability and climate resilience. This chapter investigates how blockchain-enabled Green AI can secure data integrity, enhance transparency, and improve trust in climate intelligence systems. By integrating blockchain as a decentralized trust layer with AI-driven analytics, the chapter proposes a robust framework for verifiable and auditable climate intelligence. The chapter presents an extensive literature review, a conceptual architecture, sectoral use cases, and a UAE-focused case study aligned with the UAE Net Zero 2050 Strategy.

Climate change is a threat that is being faced by us and it is a reality that this is a man-made phenomenon. Studies have shown that human activities after the 1850s have increased emission of carbon gases into the atmosphere leading to increased global atmospheric temperatures and this impacting sustainable development on a large scale. A solution to this disaster should also be provided by us and it lies in creating policies and practices that will lead to reduced carbon emissions and atmospheric temperature. Artificial intelligence techniques can analyse big data that will help build models to predict changes in the climatic patterns. This study uses traditional, machine learning and deep learning models to predict temperature based on a data set from the year 1850 to 2013. The findings suggest that XG boost and LightGBM are better than other models in terms of prediction accuracy. This indicates that AI models are critical in terms of predicting climate change patterns and in proactive monitoring and adaptation to environmental issues that affect sustainable development.

As artificial intelligence increases the computational load of models, the tacit environmental costs of predictive models emerge. This study presents a framework for quantifying and optimizing latent costs of intelligence in drug discovery. Benchmarking three methodological families physics-based approaches, machine-learning ensembles and geometric deep learning (GNN) the framework maps trade-offs between predictive accuracy, computational compactness and carbon footprint. Using resistance to kinase inhibitors as an example, GNNs provide high discriminative power (AUC = 0.89) but demand more energy, whereas ensembles offer a better accuracy–sustainability balance, requiring 0.016 kWh per inference. Despite high demands, physics-based methods retain interpretability. A sustainability scorecard combines performance, efficiency and governance indicators into an index (SI), and step-wise escalation workflows relate model complexity to decision criticality, guiding sustainable intelligence in drug discovery.

This chapter, per the authors, examines the transformative role of Artificial Intelligence (AI) in advancing climate science, weather forecasting, disaster prediction, and biodiversity monitoring. It explores the emergence of Green AI as a paradigm that balances computational efficiency, interpretability, and environmental sustainability. The chapter highlights how machine learning and deep learning models, including Pangu-Weather, GraphCast, and FuXi, have surpassed traditional numerical models in accuracy and speed. Furthermore, it discusses ethical and governance dimensions such as data transparency, algorithmic fairness, and equitable access for developing nations. By integrating responsible AI principles and sustainable computing practices, this chapter proposes a roadmap for energy-efficient, explainable, and inclusive climate intelligence that supports global resilience and aligns with the United Nations Sustainable Development Goals.

Artificial Intelligence (AI) plays a vital role in climate monitoring, biodiversity assessment, and disaster prediction; however, traditional centralized AI models demand high computational power and energy, increasing carbon emissions. This chapter introduces an energy-aware Federated Learning (FL) framework that decentralizes model training across distributed sensors to minimize communication overhead, preserve data privacy, and reduce energy consumption. By employing lightweight models, adaptive aggregation, and renewable-powered infrastructure, the framework achieves accurate, low-carbon environmental intelligence. It positions FL as a sustainable "green AI" solution for scalable, privacy-conscious, and climate-resilient monitoring systems.

The accelerating climate crisis demands solutions that surpass conventional approaches. As rising temperatures, dwindling resources, and global food insecurity intensify, artificial intelligence (AI) emerges as a transformative force capable of reshaping humanity's relationship with the planet. This chapter investigates how AI is driving sustainability across critical domains from optimizing renewable energy systems and designing resilient, low carbon cities to revolutionizing agriculture and enhancing biodiversity monitoring. In parallel, it critically examines the challenges AI introduces, including its significant energy footprint, ethical complexities, and the risk of deepening global inequities. By integrating insights from science, technology, business, and policy, this work envisions a future where intelligent systems do not merely support sustainability but embody it. Ultimately, it asks whether AI can evolve from a tool of disruption into a partner for planetary resilience and argues that the answer will shape the trajectory of both technology and life on Earth.

This chapter discusses weather forecasting and Green AI by exploring the enhancement on how energy-efficient machine learning (ML) techniques can improve weather forecast accuracy while concurrently reducing carbon and energy footprints. The objectives are to: (1) evaluate and identify modern or advance energy-efficient ML models that make weather prediction provide accuracy with lower computational cost, (2) compare eco-friendly algorithmic approaches that is suitable for resource-constrained environments, (3) integrate sustainable data processing and storage practices into forecasting systems, and (4) assess and evaluate on how Green AI in enables on effective early warning systems for extreme weather events. By analyzing model-level strategies (pruning, quantization, distillation, and efficient architectures), and discussing systems-level interventions (edge inference, optimized data pipelines, and renewable-powered infrastructures).

Agriculture faces irregular weather, intensifying natural disasters, and biodiversity loss that threaten global food security. To produce food responsibly, Agriculture 4.0 has revolutionized food production by integrating Green AI, precision technologies, and sustainable farming methods. AI-powered weather forecasting systems improve irrigation, crop scheduling, and disaster planning, while predictive models for floods, droughts, and insect outbreaks boost agricultural resilience. Additionally, remote sensing, drones, and AI algorithms track biodiversity to maintain ecosystems, enhance crop diversification, and reduce environmental consequences. IoT-based sensor networks, big data analytics, blockchain, and energy-efficient AI frameworks enable climate-smart agriculture. This chapter addresses digital inequity, data governance, and responsible AI use outside technology. Agriculture 4.0 provides a resilient, inclusive, and eco-friendly framework for feeding the globe in a changing environment by merging predictive intelligence with sustainability.

Preface

INTRODUCTION

The accelerating pace of climate change has intensified the need for innovative, scalable, and sustainable solutions capable of addressing complex environmental challenges. From extreme weather events to biodiversity loss and ecosystem degradation, the risks associated with climate change are no longer distant projections but immediate global realities. In this context, Artificial Intelligence (AI) has emerged as a transformative force, offering unprecedented capabilities in prediction, monitoring, and decision-making. However, as AI systems grow in scale and complexity, so too does their environmental footprint—raising critical questions about sustainability, accountability, and ethical deployment.

This volume, *Predicting and Monitoring Climate Risks Through Green AI: Weather Forecasting, Disaster Prediction, and Biodiversity Monitoring*, brings together cutting-edge research at the intersection of AI innovation and environmental sustainability. It introduces the concept of Green AI as a paradigm that not only leverages intelligent systems to combat climate change but also emphasizes energy efficiency, reduced carbon emissions, and responsible computation practices.

CHAPTER OVERVIEWS

The chapters in this book collectively explore how AI can be designed, optimized, and governed to serve both technological advancement and ecological preservation. They provide a multidisciplinary perspective, integrating insights from information systems, environmental science, data analytics, and sustainable development.

The volume begins with **Chapter 1**, which examines the role of blockchain-enabled Green AI in securing data integrity within climate intelligence systems—highlighting the importance of trust and transparency in environmental data ecosystems. **Chapter 2** expands on AI-driven approaches to improving predictive accuracy for sustainable development, while **Chapter 3** critically evaluates the hidden environmental costs of AI, particularly the trade-offs between computational intensity, carbon emissions, and model performance.

Chapter 4 reimagines climate science through AI applications in weather forecasting, disaster prediction, and biodiversity monitoring, setting the conceptual foundation for subsequent contributions. **Chapter 5** introduces energy-aware federated learning approaches that minimize carbon footprints in distributed environmental sensing systems, followed by **Chapter 6**, which presents a broader vision of sustainable intelligence and its role in shaping a greener and smarter planet.

Chapters 7 and 8 focus on practical applications, including eco-friendly machine learning techniques for improving weather forecast accuracy and the use of Green AI in enabling climate-resilient and sustainable agricultural practices. **Chapter 9** bridges environmental intelligence with financial systems by proposing a framework for integrating predictive climate data into risk accounting and reporting mechanisms.

Finally, **Chapter 10** addresses emerging challenges, including stakeholder trust, ethical concerns, and the risks of algorithmic greenwashing. It provides a forward-looking discussion on governance, transparency, and the responsible adoption of Green AI technologies.

CONCLUSION

Collectively, these contributions underscore a central message: technological innovation must be aligned with environmental responsibility. Green AI is not merely an optimization strategy—it is a necessary evolution in how we design, deploy, and evaluate intelligent systems in an era defined by climate urgency.

This book is intended for researchers, practitioners, policymakers, and graduate students seeking to understand and contribute to the development of sustainable AI solutions for climate risk prediction and monitoring. It is our hope that this volume will inspire further research, foster interdisciplinary collaboration, and support the global transition toward a more sustainable and resilient future.

Fanar Shwedeh
City University, Ajman, UAE

Chapter 1
Blockchain-Enabled Green AI:
Securing Data Integrity in Climate Intelligence Systems

Mohamed Al Hemairy
http://orcid.org/0000-0001-5496-0048
University of Sharjah, UAE

Ali Hilal-Alnaqbi
University of Khorfakan, UAE

Fanar Shwedeh
http://orcid.org/0009-0005-5471-8418
City University, Ajman, UAE

ABSTRACT

Blockchain and artificial intelligence (AI) are increasingly recognized as foundational technologies for advancing environmental sustainability and climate resilience. This chapter investigates how blockchain-enabled Green AI can secure data integrity, enhance transparency, and improve trust in climate intelligence systems. By integrating blockchain as a decentralized trust layer with AI-driven analytics, the chapter proposes a robust framework for verifiable and auditable climate intelligence. The chapter presents an extensive literature review, a conceptual architecture, sectoral use cases, and a UAE-focused case study aligned with the UAE Net Zero 2050 Strategy.

DOI: 10.4018/979-8-3373-7554-0.ch001

1. INTRODUCTION

Artificial intelligence (AI) has become a cornerstone of sustainability-driven innovation, supporting climate modeling, renewable energy optimization, smart city management, and biodiversity conservation (Rolnick et al., 2022; Vinuesa et al., 2020; Ahmadi et al., 2022). Governments and industries increasingly rely on AI-driven climate intelligence systems to guide policy decisions, infrastructure investments, and regulatory compliance, particularly in the context of climate risk assessment, energy transition, and environmental monitoring (European Commission, 2022; OECD, 2021; World Economic Forum, 2023a). As a result, AI-based sustainability analytics are now embedded within national climate strategies, corporate environmental governance frameworks, and international development agendas.

However, the credibility and effectiveness of these systems are fundamentally dependent on the integrity, provenance, and governance of the environmental data they process (Kshetri, 2021; Wamba et al., 2020). Climate intelligence platforms typically aggregate large volumes of heterogeneous data sourced from satellites, Internet of Things (IoT) sensors, drones, and institutional repositories. These data ecosystems are often fragmented, centrally controlled, and characterized by limited transparency, making them vulnerable to manipulation, selective reporting, and greenwashing practices (Delmas & Burbano, 2011; Lyon & Montgomery, 2015; Howson & de Vries, 2022). Such structural weaknesses undermine confidence in sustainability analytics and compromise the reliability of AI-generated insights used for environmental decision-making.

Recent studies further highlight that inconsistencies in sustainability data and environmental performance metrics have led to significant divergence in ESG assessments and climate disclosures, eroding stakeholder trust and investor confidence (Berg et al., 2022; Busch et al., 2022). These challenges underscore the urgent need for trusted digital infrastructures capable of supporting transparent, accountable, and resilient climate intelligence systems. In this context, emerging research increasingly points to the integration of advanced analytics with verifiable data governance mechanisms as a prerequisite for trustworthy sustainability-driven AI systems (Saberi et al., 2019; Kouhizadeh et al., 2021; World Economic Forum, 2023b).

2. LITERATURE REVIEW

Scholarly research on artificial intelligence (AI) for sustainability has expanded rapidly over the past decade, demonstrating the transformative potential of advanced analytics across a wide range of environmental domains. AI techniques have been successfully applied to climate modeling, extreme weather and disaster prediction, renewable energy forecasting, smart grid optimization, precision agriculture, and ecosystem and biodiversity monitoring (Rolnick et al., 2022; Vinuesa et al., 2020; Reichstein et al., 2019). Machine learning (ML) and deep learning (DL) models, in particular, have enabled the analysis of large-scale, high-dimensional environmental datasets that were previously intractable using conventional statistical methods. These advances have significantly improved the accuracy, timeliness, and scalability of climate intelligence systems used by governments, research institutions, and industry actors.

In climate science, AI-driven models are increasingly employed to enhance climate simulations, downscale global climate models, and improve the prediction of extreme weather events such as floods, heatwaves, and hurricanes (Reichstein et al., 2019; Rolnick et al., 2022). Similarly, in the energy sector, AI techniques have been widely adopted for renewable energy forecasting, load prediction, fault detection, and energy demand optimization, enabling more efficient integration of intermittent energy sources such as solar and wind into power grids (Ahmadi et al., 2022; Li et al., 2021). In agriculture and land-use management, AI supports precision farming through soil analysis, crop yield prediction, irrigation optimization, and pest detection, contributing to improved resource efficiency and food security (Kamilaris & Prenafeta-Boldú, 2018). Biodiversity monitoring has also benefited from AI-based image recognition and pattern detection applied to satellite imagery, camera traps, and drone data, allowing for large-scale and automated ecosystem assessment (Christin et al., 2019).

Despite these advances, an increasing body of literature has raised concerns regarding the environmental footprint and sustainability of AI systems themselves. Training large-scale deep learning models is computationally intensive and often associated with significant energy consumption and carbon emissions, particularly when models are trained repeatedly or deployed at scale (Strubell et al., 2019). These concerns have motivated the emergence of Green AI, a research paradigm that emphasizes energy-efficient model design, algorithmic optimization, transparency in reporting computational costs, and responsible AI deployment aligned with environmental objectives (Schwartz et al., 2020). Green AI shifts the focus from purely performance-driven benchmarks toward sustainability-aware evaluation metrics, encouraging the development of AI systems that deliver environmental benefits without exacerbating climate impacts.

Parallel to the evolution of AI for sustainability, blockchain technology has gained considerable attention as an enabling infrastructure for trustworthy and transparent environmental data management. Blockchain's core characteristics - immutability, decentralization, transparency, and cryptographic verification - have been widely discussed as mechanisms for addressing long-standing trust and governance challenges in sustainability systems (Nakamoto, 2008; Kshetri, 2021). In environmental contexts, blockchain has been applied to carbon credit markets, renewable energy certificate tracking, sustainable supply chain management, circular economy initiatives, and environmental, social, and governance (ESG) reporting platforms (Saberi et al., 2019; Kouhizadeh et al., 2021; Zhang et al., 2022).

In carbon markets, blockchain-based systems are proposed as solutions to persistent problems such as double counting, fraud, lack of transparency, and limited auditability (Zhang et al., 2022). By providing immutable transaction records and programmable smart contracts, blockchain can enhance the credibility and efficiency of emissions trading schemes and voluntary carbon markets. Similarly, in renewable energy systems, blockchain has been used to record energy generation and consumption data, enabling peer-to-peer energy trading and the transparent issuance of renewable energy certificates (Mengelkamp et al., 2018). In sustainable supply chains, blockchain facilitates traceability and provenance tracking, supporting claims related to ethical sourcing, environmental compliance, and circular economy practices (Saberi et al., 2019; Kamble et al., 2020). ESG reporting platforms increasingly explore blockchain as a means to improve the reliability and verifiability of non-financial disclosures, addressing widespread concerns regarding greenwashing and inconsistent sustainability metrics (Busch et al., 2022; Berg et al., 2022).

More recently, interdisciplinary research has begun to explore the integration of blockchain and AI as complementary technologies within sustainability-oriented digital ecosystems. In this emerging body of literature, blockchain is commonly conceptualized as a **trust layer** that secures data provenance, access control, and governance, while AI operates on verified data to generate predictive and prescriptive insights (Xu et al., 2019; Kshetri, 2021). Hybrid architectures typically store environmental data off-chain for scalability, while cryptographic hashes and metadata are recorded on-chain to ensure integrity and traceability. Smart contracts are employed to automate data validation, enforce compliance rules, and manage access rights across multiple stakeholders.

Beyond centralized learning paradigms, prior computational approaches have demonstrated the effectiveness of decentralized and rule-based intelligence models in representing complex, dynamic systems. Distributed modeling techniques, such as cellular automata, have been shown to capture emergent behaviors through localized interactions while maintaining scalability and computational efficiency. These characteristics are particularly relevant for climate intelligence systems, where heterogeneous data sources and non-linear environmental dynamics challenge monolithic AI architectures (Sgantzos et al., 2022).

Empirical and conceptual studies suggest that blockchain–AI integration can significantly enhance the trustworthiness of climate intelligence systems by reducing data manipulation, improving auditability, and supporting decentralized governance models (Rejeb et al., 2020; Kouhizadeh et al., 2021). Such integration is particularly relevant in multi-stakeholder sustainability contexts, where data is shared across public agencies, private firms, non-governmental organizations, and international bodies. By embedding trust directly into digital infrastructures, blockchain-enabled AI systems can mitigate information asymmetries and foster greater confidence in AI-generated sustainability analytics.

Despite promising results, the literature consistently identifies several unresolved challenges. From a technical perspective, scalability and interoperability remain major barriers, particularly when blockchain systems are deployed alongside data-intensive AI applications (Xu et al., 2019). Energy consumption associated with certain blockchain consensus mechanisms also raises concerns regarding their alignment with sustainability objectives, prompting growing interest in energy-efficient and permissioned blockchain architectures. From a governance standpoint, questions persist regarding data ownership, privacy, regulatory compliance, and accountability in automated decision-making systems (OECD, 2021; Kshetri, 2021). Moreover, regulatory frameworks for blockchain–AI systems remain fragmented across jurisdictions, complicating cross-border climate data sharing and global sustainability initiatives.

Overall, the literature demonstrates strong consensus on the individual value of AI and blockchain for sustainability, while highlighting the strategic potential of their integration. At the same time, it reveals critical gaps related to system-level design, governance alignment, and policy integration. These gaps underscore the need for comprehensive frameworks that combine technological innovation with regulatory, ethical, and institutional considerations. Addressing these challenges is essential for realizing the full potential of blockchain-enabled Green AI as a foundation for trustworthy and resilient climate intelligence systems.

Prior research highlights the dual role of artificial intelligence in sustainable development, demonstrating its capacity to enable a large proportion of SDG targets while simultaneously posing risks to others due to issues related to governance, energy consumption, and social equity (Vinuesa et al., 2020). This duality underscores the importance of policy-aware and responsible AI deployment in climate intelligence systems, as illustrated in Figure 1.

Figure 1. Summary of the documented positive and negative impacts of AI on the Sustainable Development Goals

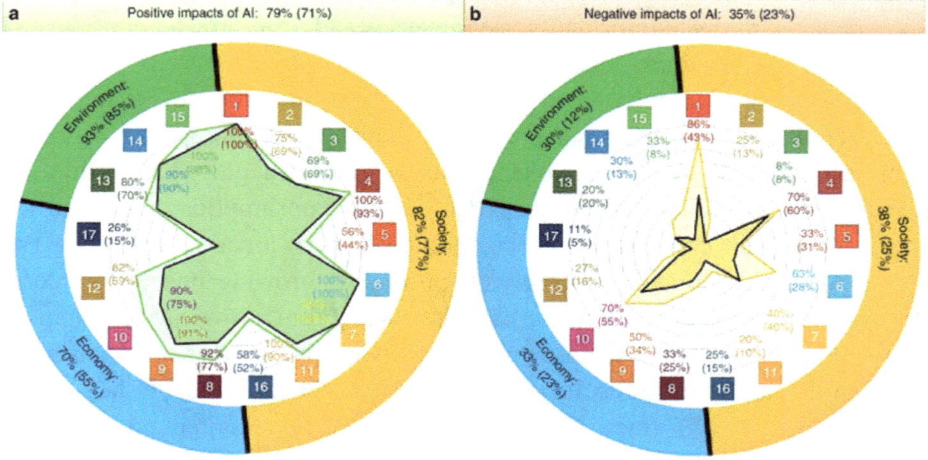

(adapted from Vinuesa et al., 2020).

Building on these insights, the next section introduces a conceptual framework for blockchain-enabled Green AI aimed at addressing trust, accountability, and policy alignment challenges in climate intelligence systems.

3. CONCEPTUAL FRAMEWORK: BLOCKCHAIN-ENABLED GREEN AI SYSTEMS

The increasing reliance on artificial intelligence (AI) for climate intelligence and sustainability analytics has intensified concerns regarding data integrity, transparency, and governance. While AI excels at extracting patterns and generating predictive insights from large environmental datasets, it does not inherently guarantee the authenticity, provenance, or immutability of the data it processes. Conversely, blockchain technology provides strong guarantees of data integrity and traceability but lacks analytical intelligence. This section proposes a blockchain-enabled Green AI conceptual framework that integrates both technologies into a unified, operational architecture for trustworthy climate intelligence systems.

3.1 Framework Rationale and Design Principles

The proposed framework is grounded in three core design principles: verifiability, scalability, and governance alignment. Verifiability ensures that all environmental data inputs and AI-generated outputs can be audited and traced back to their sources. Scalability addresses the computational and data-volume challenges inherent in climate intelligence applications. Governance alignment ensures compliance with regulatory, ethical, and institutional requirements, particularly in multi-stakeholder sustainability ecosystems (Xu et al., 2019; Kshetri, 2021).

Unlike conventional AI pipelines that rely on centralized data repositories, the proposed framework embeds trust directly into the data lifecycle. Blockchain is not used as a data storage layer for large datasets, which would be inefficient and environmentally counterproductive, but rather as a trust anchor that records cryptographic hashes, metadata, and governance rules associated with environmental data streams (Saberi et al., 2019; Kouhizadeh et al., 2021). This architectural choice aligns with Green AI principles by minimizing unnecessary computational overhead while preserving data integrity.

3.2 Multi-Layer Architecture

The framework adopts a four-layer architecture designed to support real-world deployment in climate intelligence systems.

3.2.1 Data Acquisition Layer

The data acquisition layer collects environmental data from heterogeneous sources, including Internet of Things (IoT) sensors, smart meters, satellite imagery, drones, and institutional databases. These data sources generate high-frequency, high-volume streams related to emissions levels, energy production and consumption, land use, weather patterns, and biodiversity indicators. Prior to storage or analysis, raw data undergoes preprocessing, validation, and cryptographic hashing. This step ensures that any subsequent modification of the data can be detected, thereby preserving data provenance and integrity (Rejeb et al., 2020).

3.2.2 Blockchain Trust Layer

The blockchain trust layer records hashes and metadata of environmental datasets on a permissioned or hybrid blockchain. Metadata typically includes timestamps, geolocation, data source identifiers, and access permissions. Smart contracts embedded within this layer automate data validation rules, enforce access control policies, and log data-sharing events across stakeholders (Xu et al., 2019).

A key novelty of this framework lies in its selective anchoring mechanism, whereby only critical data checkpoints and AI-relevant features are recorded on-chain. This approach reduces energy consumption and transaction costs while maintaining verifiability. It also enables interoperability with regulatory systems and ESG reporting platforms, which increasingly demand auditable sustainability data (Busch et al., 2022; Berg et al., 2022). Ensuring trust and accountability in AI-driven systems requires more than secure data storage; it necessitates verifiable records of model behavior, data provenance, and decision processes. Blockchain-based accounting mechanisms have been proposed to address this gap by enabling immutable, auditable representations of AI interactions and outputs. Such approaches extend traditional data integrity guarantees to encompass algorithmic accountability, offering a foundation for trustworthy AI governance in sustainability-critical domains (Sgantzos et al., 2023).

3.2.3 AI Analytics Layer

The AI analytics layer operates on verified, off-chain environmental data whose integrity is cryptographically anchored to the blockchain. Machine learning and deep learning models are used for tasks such as emissions forecasting, renewable energy optimization, climate risk prediction, and biodiversity assessment (Rolnick et al., 2022; Reichstein et al., 2019).

A distinctive feature of the proposed framework is the bidirectional linkage between AI outputs and blockchain records. Key AI outputs, such as aggregated emissions estimates or energy efficiency scores, are hashed and anchored to the blockchain, creating an auditable trail from raw data to decision-support indicators. This mechanism enhances accountability and enables post hoc verification of AI-driven sustainability claims, addressing growing concerns around greenwashing and opaque AI models (Delmas & Burbano, 2011; Lyon & Montgomery, 2015).

3.2.4 Application and Governance Layer

The application and governance layer provides interfaces for policymakers, regulators, industry stakeholders, and researchers. Dashboards and reporting tools present AI-generated insights alongside verifiable data provenance indicators. This layer supports regulatory compliance, ESG disclosure, and evidence-based policymaking by enabling stakeholders to trace sustainability metrics back to their underlying data sources (OECD, 2021).

Importantly, governance rules encoded within smart contracts can be aligned with institutional policies, national sustainability strategies, and international frameworks such as the United Nations Sustainable Development Goals (SDGs). This alignment ensures that technological innovation supports broader societal objectives rather than operating in isolation.

Figure 2. Conceptual Framework for Blockchain-Enabled Green AI in Climate Governance

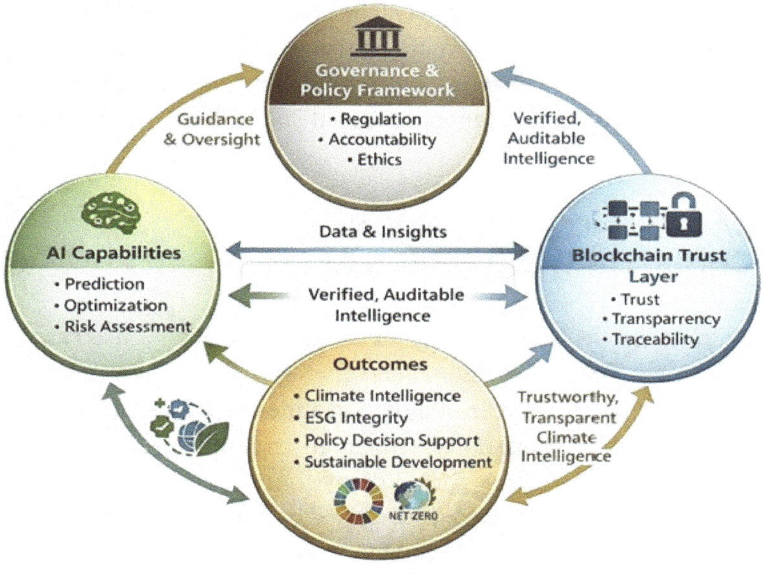

Figure 2 illustrates the integration of artificial intelligence and blockchain technologies within a governance-oriented structure to enable trustworthy, transparent, and policy-aligned climate intelligence for sustainable development.

3.3 Novel Contributions and Practical Applications

The primary novelty of this conceptual framework lies in its integration of Green AI principles with blockchain-based governance, resulting in a system that is both analytically powerful and institutionally trustworthy. Unlike prior approaches that treat blockchain and AI as loosely coupled technologies, this framework establishes a continuous trust chain spanning data acquisition, analytics, and decision-making.

In practical terms, the framework can be applied to several real-world scenarios. In smart energy systems, AI models can optimize energy distribution and predict demand, while blockchain secures generation and consumption data, enabling transparent energy trading and renewable certificate issuance (Li et al., 2021; Mengelkamp et al., 2018). In climate finance and carbon markets, the framework supports verifiable emissions tracking and reduces fraud through immutable transaction records (Zhang et al., 2022). In biodiversity conservation, AI-driven ecosystem assessments anchored to blockchain records can improve the credibility of conservation funding and environmental impact assessments (Christin et al., 2019).

Within the UAE context, such a framework aligns closely with national digital transformation and sustainability agendas. Smart city initiatives, renewable energy projects, and ESG reporting platforms can benefit from blockchain-enabled Green AI by enhancing transparency, reducing greenwashing risks, and strengthening investor and public trust. The framework is therefore not only conceptually novel but also deployable within existing institutional and technological infrastructures.

4. USE CASES AND APPLICATIONS OF BLOCKCHAIN-ENABLED GREEN AI

Blockchain-enabled systems enable sustainability-oriented use cases by transforming traditionally linear and fragmented operational structures into transparent, multi-actor ecosystems. Prior research demonstrates how blockchain technologies facilitate traceability, accountability, and real-time information sharing across complex value chains, thereby supporting environmental, social, and governance objectives. Figure 3 illustrates this transformation by contrasting conventional supply chain architectures with blockchain-enabled networks that integrate registrars, certifiers, standards organizations, and end users within a shared and auditable digital infrastructure (Saberi et al., 2019).

Figure 3. Transformation from a traditional linear supply chain to a blockchain-enabled ecosystem

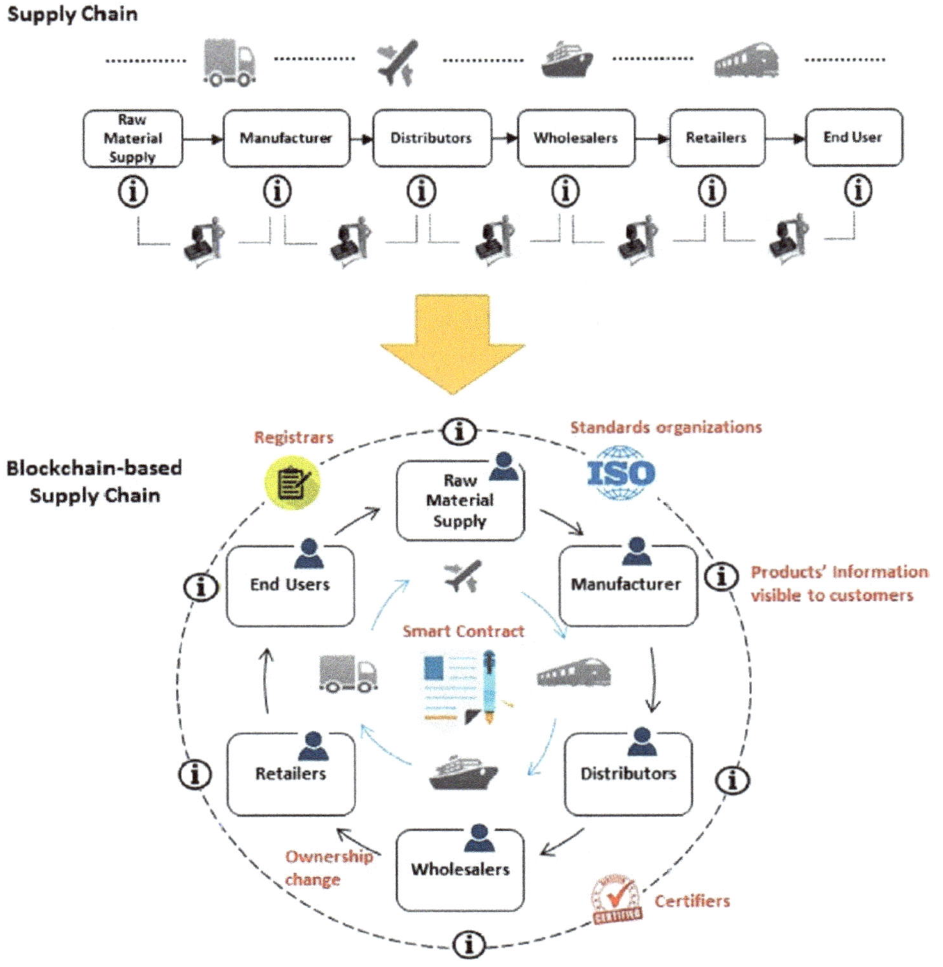

(adapted from Saberi et al., 2019)

The integration of blockchain and Green AI enables a new class of sustainability applications in which analytical intelligence is combined with verifiable, auditable, and governance-ready data infrastructures. This section examines four major application domains: smart cities, renewable energy systems, agriculture, and biodiversity monitoring, where blockchain-enabled Green AI can deliver tangible environmental, economic, and governance benefits. These use cases are selected based on their maturity, policy relevance, and demonstrated reliance on trusted environmental data.

4.1 Smart Cities and Urban Sustainability

Smart cities represent one of the most prominent application domains for AI-driven sustainability analytics. AI is widely used in urban environments to optimize traffic flows, manage energy consumption, improve waste collection, monitor air quality, and support water resource management (Batty et al., 2012; Bibri & Krogstie, 2017). Machine learning models analyze data from distributed sensors, cameras, and smart meters to improve operational efficiency and reduce environmental footprints. However, urban data infrastructures are often fragmented across municipal departments and private service providers, creating challenges related to data ownership, transparency, and accountability.

Blockchain-enabled Green AI addresses these challenges by introducing a trusted data governance layer that secures urban environmental data at the point of generation. Sensor data related to emissions, energy use, and mobility patterns can be cryptographically anchored to a blockchain, ensuring provenance and preventing post hoc manipulation (Rejeb et al., 2020). AI models operating on this verified data can then generate optimization strategies and predictive insights that are auditable by regulators and city administrators.

Practical implementations include AI-based traffic management systems whose performance metrics and emissions impact are verifiably recorded, enabling transparent evaluation of policy interventions. Similarly, waste-to-energy initiatives and smart water grids can leverage blockchain-secured data to support compliance reporting and public accountability. The integration of blockchain thus strengthens citizen trust in AI-driven urban sustainability initiatives while reducing the risk of data misuse or selective disclosure (Kshetri, 2021).

4.2 Renewable Energy Systems and Smart Grids

Renewable energy integration presents significant operational challenges due to the intermittency and decentralization of energy sources such as solar and wind. AI techniques are widely employed to forecast energy generation, predict demand, detect faults, and optimize grid stability (Li et al., 2021; Ahmad et al., 2021). These applications depend on continuous, high-quality data streams from smart meters, weather sensors, and energy management systems.

Blockchain-enabled Green AI enhances renewable energy systems by securing energy generation and consumption data, thereby enabling transparent energy transactions and reliable certification mechanisms. Blockchain-based platforms have been proposed and piloted for peer-to-peer energy trading, where prosumers exchange surplus renewable energy within local microgrids (Mengelkamp et al., 2018). In such systems, AI optimizes pricing, demand response, and load balancing, while blockchain ensures that energy transactions and renewable energy certificates are immutable and verifiable.

This integration is particularly relevant for national energy strategies and carbon accounting frameworks, where the credibility of renewable energy claims is critical. By anchoring AI-generated forecasts and performance indicators to blockchain records, stakeholders can audit energy efficiency improvements and emissions reductions with greater confidence. Moreover, permissioned blockchain architectures allow utilities and regulators to enforce compliance rules without compromising scalability or data privacy (Kouhizadeh et al., 2021).

4.3 Agriculture 4.0 and Sustainable Food Systems

Agriculture is both highly vulnerable to climate change and a significant contributor to environmental degradation. AI-driven precision agriculture has emerged as a key enabler of sustainable food systems by optimizing irrigation, fertilizer use, pest management, and crop yields (Kamilaris & Prenafeta-Boldú, 2018). Machine learning models process data from soil sensors, satellite imagery, drones, and weather forecasts to support data-driven farming decisions.

However, sustainability claims in agriculture - such as reduced emissions, organic production, or ethical sourcing - are often difficult to verify, particularly across complex supply chains. Blockchain-enabled Green AI addresses this limitation by linking AI-based agricultural analytics with immutable records of production practices, inputs, and outputs. Blockchain has been widely studied as a mechanism for improving traceability and transparency in agri-food supply chains, supporting sustainability certification and regulatory compliance (Kamble et al., 2020; Saberi et al., 2019).

In practical deployments, AI models can estimate resource efficiency and environmental impact at the farm level, while blockchain records these metrics in a tamper-resistant manner. This enables downstream actors, including distributors, retailers, and regulators, to verify sustainability claims without relying solely on self-reported data. Such systems also support fairer market access for smallholder farmers by reducing information asymmetries and enhancing trust in digital certification schemes.

4.4 Biodiversity Monitoring and Ecosystem Management

Biodiversity loss represents a critical global sustainability challenge, yet monitoring ecosystems at scale remains resource-intensive and fragmented. AI has significantly advanced biodiversity monitoring through automated species recognition, habitat mapping, and ecosystem change detection using satellite imagery, camera traps, and acoustic sensors (Christin et al., 2019; Wäldchen & Mäder, 2018). These tools enable continuous, large-scale monitoring that would be infeasible through manual methods alone.

Despite these advances, conservation funding, impact assessments, and policy interventions often rely on data whose provenance and integrity are difficult to verify. Blockchain-enabled Green AI introduces a mechanism for securing biodiversity data across the monitoring lifecycle. AI-generated indicators, such as species abundance estimates or habitat degradation scores, can be cryptographically anchored to blockchain records, ensuring that conservation outcomes are transparent and auditable.

This capability is particularly relevant in contexts where biodiversity metrics are linked to financial instruments, such as conservation credits or results-based funding mechanisms. By embedding trust into biodiversity data infrastructures, blockchain-enabled AI systems can reduce disputes, improve accountability, and strengthen confidence among donors, policymakers, and local communities (Howson & de Vries, 2022).

4.5 AI-Enabled Marine Conservation and Sustainable Aquaculture in Sharjah

Marine and aquaculture ecosystems present complex sustainability and governance challenges due to their ecological sensitivity, biological variability, and the inherent difficulty of continuous monitoring in dynamic aquatic environments. Traditional approaches to fish farming and marine resource management have typically relied on periodic sampling, manual observation, and static reporting mechanisms. Such methods limit the capacity to detect early environmental stressors, respond adaptively to ecosystem changes, and enforce sustainability standards in real time. These limitations closely parallel broader biodiversity monitoring challenges, where fragmented datasets and delayed feedback loops constrain science-based ecosystem management and conservation decision-making (Christin et al., 2019; Wäldchen & Mäder, 2018).

Artificial intelligence increasingly addresses these constraints by enabling continuous, data-driven monitoring and decision support in aquaculture and marine systems. In Sharjah, AI-enabled marine and aquaculture initiatives can be situated within a structured institutional framework that combines scientific research capabilities with regulatory oversight, notably through the Sharjah Fish Resources Authority, which is mandated to regulate fisheries, support aquaculture development, protect marine biodiversity, and enhance food security through sustainable practices (Sharjah Fish Resources Authority, 2025). Within such contexts, AI-based systems support real-time analysis of water quality parameters (e.g., temperature, dissolved oxygen, salinity), feeding efficiency, growth patterns, and fish health indicators.

Machine learning models are increasingly applied to optimize feeding regimes, reduce resource waste, and detect early signs of disease or environmental stress, improving production efficiency while mitigating ecological impacts (Kamilaris & Prenafeta-Boldú, 2018; Capetillo-Contreras et al., 2024; Yang et al., 2025).

Beyond operational optimization, AI-enabled aquaculture systems contribute to sustainability governance by strengthening monitoring continuity, supporting evidence-based regulatory compliance, and enabling science-driven resource management. When embedded within institutional and regulatory frameworks, AI systems facilitate the alignment of production practices with environmental protection objectives and biodiversity conservation goals. Continuous data collection and analytics support traceable decision-making, allowing authorities to assess compliance with sustainability standards and respond proactively to emerging ecological risks. This governance-oriented application of AI reinforces accountability and transparency in marine resource management, addressing long-standing concerns related to overexploitation, habitat degradation, and data opacity.

Extending AI-enabled sustainability frameworks from terrestrial ecosystems to marine and aquaculture environments underscores the broader role of artificial intelligence in biodiversity monitoring and ecosystem stewardship. By supporting adaptive management, food security strategies, and marine conservation objectives, AI-driven aquaculture systems complement climate intelligence and environmental governance mechanisms discussed throughout this section. This extension highlights the importance of integrating advanced analytics with institutional oversight to ensure that technological innovation contributes meaningfully to long-term ecological resilience and sustainable development.

4.6 Cross-Sectoral Value and Institutional Relevance

Across these use cases, the value of blockchain-enabled Green AI lies not merely in technological sophistication, but in its ability to support institutional trust, regulatory compliance, and multi-stakeholder coordination. Sustainability initiatives increasingly operate across organizational and jurisdictional boundaries, requiring shared data infrastructures that balance transparency with governance control (OECD, 2021).

By combining AI's analytical power with blockchain's trust mechanisms, the proposed applications address persistent challenges related to data credibility, greenwashing, and accountability. Importantly, these systems are deployable using existing technologies and governance frameworks, making them suitable for gradual integration into national sustainability strategies, ESG reporting platforms, and climate policy instruments.

5. APPLIED CASE FROM THE UAE: BLOCKCHAIN-ENABLED GREEN AI IN PRACTICE

The United Arab Emirates (UAE) provides a compelling real-world context for examining the deployment of blockchain-enabled Green AI within national sustainability and digital transformation agendas. Over the past decade, the UAE has positioned itself as a regional and global leader in renewable energy, smart city development, and climate governance, supported by large-scale investments in digital infrastructure and advanced technologies. Central to this strategy is the integration of artificial intelligence (AI), data analytics, and emerging trust technologies to support evidence-based sustainability decision-making and transparent environmental governance (UAE Government, 2021; World Economic Forum, 2023a).

5.1 Masdar City as a Living Laboratory for Green AI

Masdar City, located in Abu Dhabi, represents one of the world's most ambitious planned eco-cities and functions as a living laboratory for sustainable urban development. Designed to minimize environmental impact while maximizing energy efficiency, Masdar integrates renewable energy generation, smart buildings, and data-driven urban management systems (Reiche, 2010). AI plays a central role in optimizing energy consumption, managing building performance, forecasting renewable energy output, and supporting mobility and logistics systems within the city (IEA, 2023).

AI-driven energy management systems deployed at Masdar rely on continuous data streams from smart meters, sensors, and building management platforms. These systems use machine learning models to predict demand patterns, optimize energy distribution, and reduce peak loads, thereby improving overall system efficiency. However, as in other smart city contexts, the effectiveness and credibility of these AI applications depend on the integrity and governance of the underlying data. Blockchain-enabled Green AI provides a mechanism for anchoring critical energy and environmental data to tamper-resistant ledgers, ensuring traceability and auditability across the energy lifecycle (Kshetri, 2021; Kouhizadeh et al., 2021).

5.2 Smart Grids and Renewable Energy Governance

The UAE's national energy transition strategy places strong emphasis on renewable energy integration and grid modernization. Large-scale solar projects, such as Noor Abu Dhabi and Mohammed bin Rashid Al Maktoum Solar Park, generate vast volumes of operational data related to energy production, grid stability, and emissions reductions. AI techniques are widely used to forecast solar output, manage intermittency, detect anomalies, and optimize grid operations (Li et al., 2021; Ahmad et al., 2021).

Blockchain-enabled Green AI enhances these smart grid systems by providing trusted data governance mechanisms that support transparent energy accounting and regulatory oversight. Blockchain can be used to record renewable energy generation data, issue verifiable renewable energy certificates, and track energy flows across decentralized microgrids. AI models operating on blockchain-anchored data can then generate forecasts and optimization strategies that are verifiable by utilities, regulators, and investors (Mengelkamp et al., 2018).

In the UAE context, such systems align closely with regulatory requirements and national climate commitments. By embedding trust directly into digital energy infrastructures, blockchain-enabled Green AI reduces disputes over energy attribution and emissions reporting, strengthens confidence in renewable energy claims, and supports cross-sector coordination between energy producers, grid operators, and policymakers.

5.3 ESG Reporting and Anti-Greenwashing Mechanisms

Environmental, social, and governance (ESG) reporting has become a critical component of corporate sustainability and investment decision-making in the UAE, particularly as the country seeks to attract international capital aligned with climate and sustainability goals. However, global research highlights significant inconsistencies and credibility challenges in ESG data, including divergent ratings, selective disclosure, and greenwashing practices (Berg et al., 2022; Busch et al., 2022).

Blockchain-enabled Green AI offers a robust solution to these challenges by strengthening the verifiability of ESG metrics. In a UAE deployment scenario, AI systems can aggregate and analyze environmental performance indicators - such as energy efficiency, emissions intensity, and resource use-across corporate operations. Blockchain can then be used to anchor these metrics to immutable records, creating auditable ESG data trails that regulators, investors, and auditors can independently verify.

This approach is particularly relevant in regulated sectors such as energy, real estate, and infrastructure, where ESG disclosures increasingly influence licensing, financing, and public trust. By reducing reliance on self-reported data and enhancing transparency, blockchain-enabled Green AI directly addresses greenwashing risks and aligns ESG reporting practices with international standards and investor expectations (Delmas & Burbano, 2011; Lyon & Montgomery, 2015).

5.5 Alignment with UAE Net Zero 2050 Strategy

The UAE Net Zero 2050 Strategy represents a comprehensive national commitment to achieving climate neutrality while sustaining economic growth. The strategy emphasizes innovation, digital transformation, and public-private collaboration as key enablers of decarbonization across energy, industry, transport, and urban systems (UAE Government, 2021). Blockchain-enabled Green AI aligns closely with these priorities by providing scalable, data-driven infrastructures for monitoring progress, evaluating policy effectiveness, and supporting adaptive governance.

Within this strategic context, the proposed framework supports continuous measurement and verification of emissions reductions, renewable energy deployment, and sustainability performance. AI-driven analytics enable predictive insights and optimization, while blockchain ensures that reported outcomes are transparent, consistent, and auditable over time. This combination strengthens institutional trust and enhances the UAE's credibility in international climate forums and sustainability rankings.

5.4 Transferability and Regional Implications

Beyond its national relevance, the UAE case offers transferable lessons for other Gulf Cooperation Council (GCC) countries and emerging economies pursuing sustainability transitions under data-intensive conditions. The UAE's experience demonstrates that blockchain-enabled Green AI systems can be deployed incrementally, integrated with existing infrastructure, and governed through hybrid public-private models. These characteristics make the approach particularly suitable for regions seeking to balance rapid digitalization with environmental accountability and regulatory oversight (OECD, 2021).

Overall, the UAE mini-case study illustrates how blockchain-enabled Green AI can move beyond conceptual promise to practical implementation. By combining AI-driven intelligence with verifiable data governance, the UAE demonstrates a realistic pathway for deploying trustworthy climate intelligence systems at scale.

6. CHALLENGES AND POLICY CONSIDERATIONS

While blockchain-enabled Green AI presents a promising pathway for trustworthy and transparent climate intelligence, its large-scale deployment raises a range of technical, regulatory, ethical, and institutional challenges. Addressing these challenges is essential to ensure that such systems not only deliver analytical value but also align with sustainability objectives, legal frameworks, and public trust expectations. This section critically examines these issues, with particular attention to policy relevance and governance implications.

6.1 Technical Challenges: Scalability, Interoperability, and Energy Efficiency

From a technical perspective, one of the most frequently cited challenges in blockchain-AI integration is scalability. Climate intelligence systems generate vast volumes of high-frequency data from IoT sensors, satellites, and monitoring platforms. Blockchain systems, particularly those based on public or energy-intensive consensus mechanisms, are not designed to store or process such data at scale (Xu et al., 2019). As a result, most practical implementations rely on hybrid architectures in which data is stored off-chain and only cryptographic hashes and metadata are recorded on-chain.

Interoperability represents an additional barrier, as climate intelligence ecosystems typically involve heterogeneous data standards, legacy systems, and multiple institutional actors. The absence of common protocols for environmental data exchange complicates integration across sectors and jurisdictions (OECD, 2021). Without standardization, blockchain-enabled Green AI systems risk becoming siloed solutions that undermine their potential for cross-border climate governance.

Energy efficiency is another critical concern. While blockchain is often promoted as a trust-enabling technology for sustainability, certain consensus mechanisms have been criticized for their high energy consumption, which may conflict with environmental objectives (Howson & de Vries, 2022). This tension has prompted growing interest in permissioned blockchains, proof-of-stake mechanisms, and low-energy distributed ledger technologies that better align with Green AI principles (Kouhizadeh et al., 2021).

6.2 Data Governance, Privacy, and Regulatory Compliance

Blockchain-enabled Green AI systems operate at the intersection of data governance and automated decision-making, raising complex legal and regulatory questions. Environmental data often includes geospatial information, operational metrics, and proprietary business data, which may be subject to privacy, confidentiality, and national security regulations. Immutable blockchain records can conflict with legal requirements related to data modification, deletion, or correction, particularly under data protection regimes such as the GDPR (OECD, 2021). The environmental footprint of large-scale AI deployments has intensified concerns regarding the sustainability of data-intensive and compute-heavy models. Prior research demonstrates that efficiency-driven algorithmic design can substantially reduce computational overhead while maintaining high predictive performance, particularly in real-time and resource-constrained environments. These findings reinforce the importance of Green AI principles when designing climate intelligence systems intended for continuous operation at scale (Ahmed et al., 2018).

Furthermore, climate intelligence systems increasingly support regulatory compliance and policy enforcement, which amplifies the importance of data accuracy and accountability. Errors or biases embedded in AI models may lead to incorrect emissions estimates, flawed sustainability assessments, or unjust regulatory outcomes. Policymakers must therefore consider how liability is allocated when automated systems contribute to environmental decision-making (Kshetri, 2021).

To address these concerns, governance frameworks must clearly define data ownership, access rights, and responsibility across all stakeholders involved. Blockchain can support such frameworks by encoding access controls and audit trails, but it does not eliminate the need for legal clarity and institutional oversight.

6.3 Ethical Challenges and Algorithmic Accountability

Ethical considerations play a central role in the deployment of AI for sustainability. Although AI is often presented as an objective and data-driven tool, numerous studies demonstrate that AI systems can reproduce or amplify biases present in training data or system design (Vinuesa et al., 2020). In the context of climate intelligence, such biases may disproportionately affect vulnerable communities, small producers, or developing regions.

Blockchain-enabled Green AI introduces opportunities for improved accountability by enabling traceability of data inputs and AI outputs. However, transparency at the data level does not automatically translate into transparency at the algorithmic level. Many AI models used in climate analytics remain complex and opaque, limiting the ability of stakeholders to understand or challenge their outputs (Wamba et al., 2020). The practical deployment of blockchain-enabled Green AI systems requires modular architectures that support scalability, interoperability, and governance-by-design. Prior work on large-scale, cost-effective monitoring systems highlights the value of layered architectures that decouple data acquisition, analytics, and application interfaces, thereby enabling flexible deployment across heterogeneous environments. These architectural principles inform the proposed implementation roadmap for climate intelligence systems operating across institutional and sectoral boundaries (Al Hemairy et al., 2016).

Ethical deployment therefore requires complementary measures, including explainable AI techniques, independent audits, and participatory governance mechanisms. Without these safeguards, blockchain-enabled systems risk reinforcing technocratic decision-making that excludes affected stakeholders from meaningful participation.

6.4 Greenwashing, ESG Governance, and Market Integrity

One of the most significant policy drivers behind blockchain-enabled Green AI is the growing concern over greenwashing and unreliable sustainability claims. Research consistently highlights inconsistencies in ESG ratings, selective disclosure practices, and the limited verifiability of corporate environmental data (Delmas & Burbano, 2011; Berg et al., 2022). These challenges undermine market confidence and weaken the effectiveness of sustainability-oriented regulation.

Blockchain-enabled Green AI offers a mechanism to strengthen ESG governance by anchoring sustainability metrics to verifiable data trails. However, the effectiveness of this approach depends on regulatory recognition and standardization. Without alignment between technological systems and regulatory frameworks, blockchain-based ESG data may remain underutilized or contested in legal and financial contexts (Busch et al., 2022).

Policymakers therefore face the challenge of integrating digital verification mechanisms into existing reporting and compliance regimes, ensuring that technological innovation translates into enforceable standards rather than parallel, voluntary systems.

6.5 Institutional and Policy Coordination Challenges

Climate intelligence systems operate across institutional boundaries, involving government agencies, private firms, financial institutions, and civil society organizations. Blockchain-enabled Green AI requires coordinated governance structures capable of managing shared infrastructures, resolving disputes, and adapting to evolving regulatory landscapes (World Economic Forum, 2023b).

In the absence of clear institutional leadership, such systems may suffer from fragmented implementation, uneven adoption, and conflicting incentives. National sustainability strategies, such as the UAE Net Zero 2050 initiative, illustrate the importance of aligning digital innovation with long-term policy objectives and public-private collaboration frameworks. However, replicating such alignment across jurisdictions remains a significant challenge, particularly in cross-border climate governance contexts.

6.6 Policy Implications and Design Considerations

The challenges outlined above highlight the need for policy-aware system design. Blockchain-enabled Green AI should not be treated solely as a technical solution, but as a socio-technical system embedded within legal, ethical, and institutional environments. Effective policy responses may include:

- Development of standards for environmental data integrity and AI accountability
- Regulatory sandboxes to test blockchain-AI applications in controlled settings
- Incentives for adopting low-energy and interoperable technologies
- Integration of digital verification mechanisms into ESG and climate reporting frameworks

By addressing these considerations proactively, policymakers and system designers can ensure that blockchain-enabled Green AI contributes meaningfully to sustainable development rather than introducing new risks or inefficiencies.

7. PROPOSED MODEL AND IMPLEMENTATION ROADMAP

Building on the challenges and policy considerations discussed in Section 6, this section proposes an integrated model for deploying blockchain-enabled Green AI in climate intelligence systems. The model is designed as a modular, policy-aware architecture that balances data integrity, analytical performance, regulatory compliance, and sustainability objectives. Rather than treating blockchain and AI as isolated technologies, the proposed approach conceptualizes them as complementary components within a socio-technical system for climate governance.

7.1 Conceptual Architecture of Blockchain-Enabled Green AI

The proposed model consists of five interdependent layers:

7.1.1 Data Acquisition Layer

The model begins with the Data Acquisition Layer, where environmental data is collected from heterogeneous sources such as IoT sensors, satellite imagery, smart meters, and institutional databases. This data is preprocessed and validated before entering the analytical pipeline.

7.1.2 AI Analytics Layer

The AI Analytics Layer applies machine learning and predictive models to generate climate intelligence outputs, including emissions forecasts, renewable energy optimization insights, and environmental risk assessments. Green AI principles guide model selection and training to minimize computational and energy overhead.

7.1.3 Blockchain Trust Layer

The Blockchain Trust Layer records cryptographic hashes, timestamps, and metadata associated with data inputs and AI outputs on a distributed ledger. This approach enables verifiable data provenance and auditability without storing large datasets on-chain.

7.1.4 Governance and Access Control Layer

The Governance and Access Control Layer uses smart contracts to define data access rights, compliance rules, and stakeholder responsibilities. This layer supports regulatory oversight and controlled transparency across public and private actors.

7.1.5 Application and Policy Interface Layer

Finally, the Application and Policy Interface Layer translates verified AI outputs into actionable insights through dashboards, reporting tools, and ESG disclosure systems, enabling informed decision-making by policymakers, regulators, and market participants.

Figure 4 illustrates the proposed blockchain-enabled Green AI model for climate intelligence systems. The architecture is structured as a layered framework that integrates artificial intelligence analytics with blockchain-based trust mechanisms to ensure data integrity, transparency, and policy alignment. This layered architecture enables interoperability, modular adoption, and alignment with evolving policy and technological standards.

Figure 4. Blockchain-Enabled Green AI Architecture for Climate Intelligence

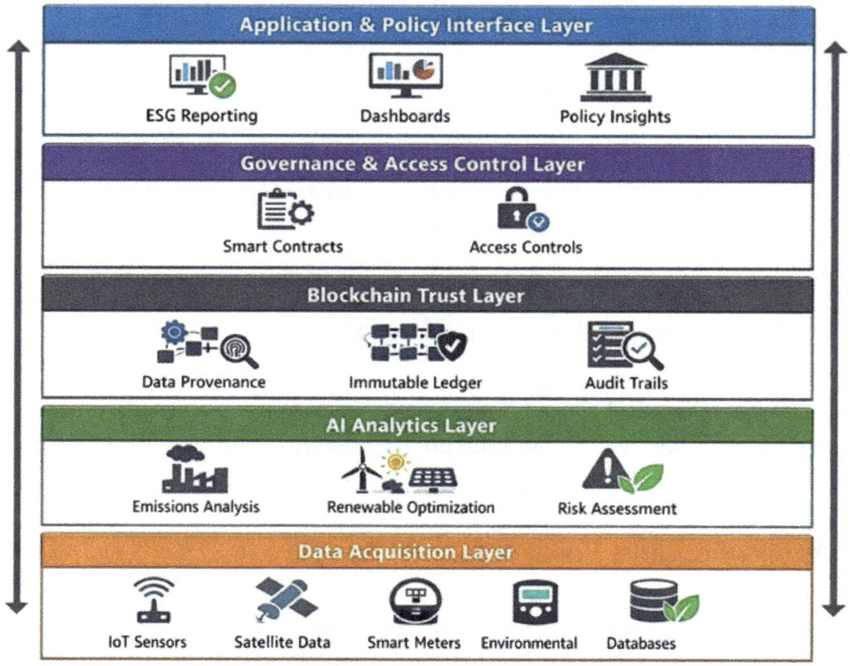

7.2 Implementation Roadmap

To operationalize the proposed model, a phased implementation roadmap is recommended.

Phase 1: Pilot Design and Stakeholder Alignment

Initial implementation should focus on clearly defined pilot projects, such as monitoring emissions within a specific industrial sector or renewable energy optimization within a smart-city district. Stakeholder roles, governance structures, and data ownership arrangements must be established at this stage.

Phase 2: Technical Integration and Validation

During this phase, AI models and blockchain infrastructure are integrated using hybrid architectures. Performance, scalability, and energy-efficiency metrics are evaluated to ensure alignment with Green AI objectives. Regulatory and compliance requirements should be tested through sandbox environments.

Phase 3: Policy Integration and Standardization

Successful pilots can be scaled through alignment with national sustainability strategies and reporting frameworks. Integration with ESG disclosure requirements and climate reporting standards enhances institutional adoption and policy relevance.

Phase 4: Scaling and Cross-Sector Expansion

The final phase involves extending the system across sectors and jurisdictions. Interoperability standards and governance mechanisms must be refined to support cross-border climate intelligence and multi-stakeholder collaboration.

7.3 Alignment with Sustainability and Climate Policy Objectives

The proposed model directly supports global and national sustainability goals by enhancing transparency, accountability, and data-driven decision-making. By enabling verifiable climate intelligence, blockchain-enabled Green AI can strengthen progress toward the UN Sustainable Development Goals, particularly those related to climate action, responsible production, and sustainable infrastructure.

National strategies, such as the UAE Net Zero 2050 initiative, provide a policy context in which such systems can be deployed as part of broader digital transformation agendas. Aligning technical design with policy objectives ensures that innovation contributes to long-term environmental and economic resilience rather than fragmented technological experimentation.

7.4 Practical Implications for Policymakers and System Designers

For policymakers, the proposed model highlights the importance of embedding digital verification mechanisms within existing regulatory frameworks. For system designers, it emphasizes the need for energy-aware AI development, interoperable blockchain architectures, and governance-by-design approaches.

Ultimately, the successful deployment of blockchain-enabled Green AI depends not only on technological maturity but also on institutional coordination, regulatory clarity, and sustained stakeholder engagement.

8. CONCLUSION AND FUTURE DIRECTIONS

This chapter examined the emerging role of blockchain-enabled Green AI as a foundational infrastructure for trustworthy and transparent climate intelligence systems. By integrating artificial intelligence with distributed ledger technologies, the chapter addressed a critical gap in sustainability analytics - namely, the need for verifiable, auditable, and accountable environmental data in support of climate governance, regulatory compliance, and market integrity. Rather than positioning technology as a standalone solution, the analysis emphasized the socio-technical nature of climate intelligence and the importance of aligning innovation with policy, ethics, and institutional frameworks.

Through a structured review of existing literature, the chapter highlighted how Green AI can enhance predictive capabilities across domains such as renewable energy management, emissions monitoring, smart cities, and biodiversity assessment. At the same time, it demonstrated that blockchain technologies can strengthen trust in these systems by securing data provenance, enabling traceability, and mitigating risks associated with manipulation, greenwashing, and fragmented governance. The proposed model and implementation roadmap illustrated how these complementary technologies can be deployed in a scalable and policy-aware manner, balancing analytical performance with sustainability and regulatory objectives.

Looking forward, several directions for future research and practice emerge. From a technical perspective, further work is needed to develop energy-efficient AI models and interoperable blockchain architectures tailored to large-scale climate data ecosystems. Advances in explainable AI and privacy-preserving computation will be essential to enhance transparency and regulatory acceptance. From a governance standpoint, comparative studies across jurisdictions can shed light on how different regulatory approaches influence the adoption and effectiveness of blockchain-enabled climate intelligence. In addition, empirical evaluations of pilot projects will be critical to assess real-world impacts, cost-effectiveness, and social implications.

Ultimately, the success of blockchain-enabled Green AI will depend on sustained collaboration between policymakers, technologists, industry actors, and civil society. When designed and governed responsibly, such systems have the potential to transform climate intelligence from fragmented and opaque data practices into a robust digital public infrastructure that supports informed decision-making and long-term sustainable development.

REFERENCES

Ahmad, T., Zhang, D., Huang, C., Zhang, H., Dai, N., Song, Y., & Chen, H. (2021). Artificial intelligence in sustainable energy industry: Status quo, challenges and opportunities. *Journal of Cleaner Production, 289*, 125834. DOI: 10.1016/j.jclepro.2021.125834

Ahmadi, M. H., Ghazvini, M., Sadeghzadeh, M., Nazari, M. A., Kumar, R., Naeimi, A., & Ming, T. (2022). Application of artificial intelligence for renewable energy systems: A review. *Renewable & Sustainable Energy Reviews, 156*, 111821.

Ahmed, S., Hilal-Alnaqbi, A., Al Hemairy, M., & Al Ahmad, M. (2018). ECG abnormality detection algorithm. *International Journal of Advanced Computer Science and Applications, 9*(8), 215–219. DOI: 10.14569/IJACSA.2018.090827

Al Hemairy, M., Amin, S., Hijji, M., Serhani, M., & Al Ahmad, M. (2016). Integrated and scalable architecture for providing cost-effective remote health monitoring. In *9th International Conference on Developments in eSystems Engineering (DeSE)* (pp. 74–80). IEEE. DOI: 10.1109/DeSE.2016.14

Batty, M., Axhausen, K. W., Giannotti, F., Pozdnoukhov, A., Bazzani, A., Wachowicz, M., Ouzounis, G., & Portugali, Y. (2012). Smart cities of the future. *The European Physical Journal. Special Topics, 214*(1), 481–518. DOI: 10.1140/epjst/e2012-01703-3

Berg, F., Koelbel, J. F., & Rigobon, R. (2022). Aggregate confusion: The divergence of ESG ratings. *Review of Finance, 26*(6), 1315–1344. DOI: 10.1093/rof/rfac033

Bibri, S. E., & Krogstie, J. (2017). Smart Sustainable Cities of the Future: An Extensive Interdisciplinary Literature Review. *Sustainable Cities and Society, 31*, 183–212. DOI: 10.1016/j.scs.2017.02.016

Busch, T., Johnson, M., & Pioch, T. (2022). Corporate carbon performance data: Quo vadis? *Journal of Industrial Ecology, 26*(1), 350–363. DOI: 10.1111/jiec.13008

Capetillo-Contreras, O., Rodríguez-Pulido, J. A., Hernández-Aguilar, M., & Parra-Flores, J. (2024). Artificial intelligence-based aquaculture systems: Monitoring, management, and sustainability perspectives. *Journal of Marine Science and Engineering, 12*(1), 161. DOI: 10.3390/jmse12010161

Christin, S., Hervet, É., & Lecomte, N. (2019). Applications for deep learning in ecology. *Methods in Ecology and Evolution, 10*(10), 1632–1644. DOI: 10.1111/2041-210X.13256

Delmas, M. A., & Burbano, V. C. (2011). The drivers of greenwashing. *California Management Review*, *54*(1), 64–87. DOI: 10.1525/cmr.2011.54.1.64

European Commission. (2022). *Artificial intelligence and climate change*. Publications Office of the European Union.

. UAE Government. (2021). UAE Net Zero by 2050 strategic initiative.

Howson, P., & de Vries, A. (2022). Preying on the poor? Opportunities and risks of blockchain technologies for climate finance. *Energy Research & Social Science*, *84*, 102394. DOI: 10.1016/j.erss.2021.102394

International Energy Agency (IEA). (2023). *Digitalization and energy*. IEA Publishing.

Kamble, S. S., Gunasekaran, A., & Arha, H. (2020). Understanding the blockchain technology adoption in supply chains. *International Journal of Production Research*, *58*(7), 2009–2033. DOI: 10.1080/00207543.2018.1518610

Kamilaris, A., & Prenafeta-Boldú, F. X. (2018). Deep learning in agriculture: A survey. *Computers and Electronics in Agriculture*, *147*, 70–90. DOI: 10.1016/j.compag.2018.02.016

Kouhizadeh, M., Sarkis, J., & Zhu, Q. (2021). At the nexus of blockchain technology, the circular economy, and product deletion. *Transportation Research Part E, Logistics and Transportation Review*, *145*, 102153.

Kshetri, N. (2021). Artificial intelligence, blockchain, and trust. *Computer*, *54*(1), 88–92.

Li, Z., Shahidehpour, M., Alabdulwahab, A., & Abusorrah, A. (2021). Bilevel model for power system operation with demand response and energy storage. *IEEE Transactions on Smart Grid*, *12*(4), 2921–2932.

Lyon, T. P., & Montgomery, A. W. (2015). The means and end of greenwash. *Organization & Environment*, *28*(2), 223–249. DOI: 10.1177/1086026615575332

Mengelkamp, E., Gärttner, J., Rock, K., Kessler, S., Orsini, L., & Weinhardt, C. (2018). Designing microgrid energy markets. *Applied Energy*, *210*, 870–880. DOI: 10.1016/j.apenergy.2017.06.054

Nakamoto, S. (2008). Bitcoin: A peer-to-peer electronic cash system. https://bitcoin.org/bitcoin.pdf

OECD. (2021). *Artificial intelligence, data governance and privacy: Policy considerations*. OECD Publishing.

Reiche, D. (2010). Renewable energy policies in the Gulf countries: A case study of the carbon-neutral "Masdar City" in Abu Dhabi. *Energy Policy*, *38*(1), 378–382. DOI: 10.1016/j.enpol.2009.09.028

Reichstein, M., Camps-Valls, G., Stevens, B., Jung, M., Denzler, J., Carvalhais, N., & Prabhat, . (2019). Deep learning and process understanding for data-driven Earth system science. *Nature*, *566*(7743), 195–204. DOI: 10.1038/s41586-019-0912-1 PMID: 30760912

Rejeb, A., Keogh, J. G., & Treiblmaier, H. (2020). Leveraging the Internet of Things and Blockchain Technology in Supply Chain Management. *Future Generation Computer Systems*, *107*, 501–514.

Rolnick, D., Donti, P. L., Kaack, L. H., Kochanski, K., Lacoste, A., Sankaran, K., Ross, A. S., Milojevic-Dupont, N., Jaques, N., Waldman-Brown, A., Luccioni, A., Maharaj, T., Sherwin, E. D., Mukkavilli, S. K., Kording, K. P., Gomes, C. P., Ng, A. Y., Hassabis, D., Platt, J. C., & Bengio, Y. (2022). Tackling climate change with machine learning. ACM Computing Surveys, 55(2). *Article*, *42*, 1–96.

Saberi, S., Kouhizadeh, M., Sarkis, J., & Shen, L. (2019). Blockchain technology and its relationships to sustainable supply chain management. *International Journal of Production Research*, *57*(7), 2117–2135. DOI: 10.1080/00207543.2018.1533261

Schwartz, R., Dodge, J., Smith, N. A., & Etzioni, O. (2020). Green AI. *Communications of the ACM*, *63*(12), 54–63. DOI: 10.1145/3381831

Sgantzos, K., Al Hemairy, M. A., Tzavaras, P., & Stelios, S. (2023). Triple-Entry Accounting as a Means of Auditing Large Language Models. *Journal of Risk and Financial Management*, *16*(9), 383. DOI: 10.3390/jrfm16090383

Sgantzos, K., Grigg, I., & Al Hemairy, M. (2022). Multiple Neighborhood Cellular Automata as a Mechanism for Creating an AGI on a Blockchain. *Journal of Risk and Financial Management*, *15*(8), 360. DOI: 10.3390/jrfm15080360

Sharjah Fish Resources Authority. (2025). Sharjah Fish Resources Authority. Government of Sharjah. https://fra.shj.ae/en

Strubell, E., Ganesh, A., & McCallum, A. (2019). Energy and policy considerations for deep learning in NLP. *Proceedings of ACL*, 3645-3650. DOI: 10.18653/v1/P19-1355

Vinuesa, R., Azizpour, H., Leite, I., Balaam, M., Dignum, V., Domisch, S., & Nerini, F. F. (2020). The role of artificial intelligence in achieving the Sustainable Development Goals. *Nature Communications*, *11*(1), 233. DOI: 10.1038/s41467-019-14108-y PMID: 31932590

Wäldchen, J., & Mäder, P. (2018). Machine learning for biodiversity research. *Ecological Informatics*, *45*, 88–95.

Wäldchen, J., & Mäder, P. (2018). Machine learning for image-based species identification. *Ecological Informatics*, *45*, 88–95.

Wamba, S. F., Gunasekaran, A., Akter, S., Ren, S. J., Dubey, R., & Childe, S. J. (2020). Big data analytics and firm performance: Effects of dynamic capabilities. *Journal of Business Research*, *111*, 88–100.

World Economic Forum. (2023a). Blockchain for climate action. World Economic Forum.

World Economic Forum. (2023b). Harnessing artificial intelligence for climate and sustainability. World Economic Forum.

Xu, X., Weber, I., & Staples, M. (2019). *Architecture for blockchain applications*. Springer. DOI: 10.1007/978-3-030-03035-3

Yang, H., Li, D., Zhao, X., & Chen, Y. (2025). Artificial intelligence-driven aquaculture: Applications, challenges, and sustainability implications. *Aquaculture Reports*, *27*, 101373.

Zhang, Y.. (2022). Blockchain-based carbon trading mechanisms and policy implications. *Energy Policy*, *162*, 112776.

Chapter 2

AI for Combating Climate Change:

Improving Prediction Accuracy for Sustainable Development

Harish Uppilappatta Chennelleri
http://orcid.org/0000-0003-1800-6294
City University, Ajman, UAE

N. M. Dhanya
Murdoch University, Dubai, UAE

ABSTRACT

Climate change is a threat that is being faced by us and it is a reality that this is a man-made phenomenon. Studies have shown that human activities after the 1850s have increased emission of carbon gases into the atmosphere leading to increased global atmospheric temperatures and this impacting sustainable development on a large scale. A solution to this disaster should also be provided by us and it lies in creating policies and practices that will lead to reduced carbon emissions and atmospheric temperature. Artificial intelligence techniques can analyse big data that will help build models to predict changes in the climatic patterns. This study uses traditional, machine learning and deep learning models to predict temperature based on a data set from the year 1850 to 2013. The findings suggest that XG boost and LightGBM are better than other models in terms of prediction accuracy. This indicates that AI models are critical in terms of predicting climate change patterns and in proactive monitoring and adaptation to environmental issues that affect sustainable development.

DOI: 10.4018/979-8-3373-7554-0.ch002

INTRODUCTION

Climate change is a threat that is being faced by the Earth's inhabitants and it is a reality that this is a man-made phenomenon. Studies have shown that human activities including industrial activities after the 1850s have increased emission of carbon gases into the atmosphere leading to increased global atmospheric temperatures. As per Intergovernmental Panel on Climate Change (IPCC), human induced warming reached 1^0C by the year 2017 and is increasing at the rate of 0.2^0C every decade. Climate alteration is impacting sustainable development on a large scale. Irregular weather patterns are adversely impacting food production systems, and biodiversity, directly impacting the attainment of several of the sustainable development goals. A solution to this disaster should also be provided by us and it lies in creating policies and practices that will lead to reduced carbon emissions and atmospheric temperature. This will lead to a better Earth for all its inhabitants enabling them to survive and grow. There are increasing occurrences of global warming induced extreme weather conditions in different parts of the world resulting in loss of life, and population displacements. It is also worth noting that the adverse impacts are more prominent in economies that are poorer and marginalized populations are the ones who are deeply impacted. Any attempt to mitigate these risks will help the population that do not have access to the latest economic and technological developments. We need innovative methodologies to be applied to mitigate the risk that is in front of us. Artificial intelligence techniques can enable analysis of big data that will help to build models to predict changes in the climatic patterns. Developments in machine learning languages and deep learning languages can monitor and predict weather patterns and help in taking appropriate actions to reduce the impact of disasters. These models also assist in formulating appropriate policies and developing practices that can assist in combating climate change and mitigating the associated risks. Climate change is dependent on a number of different factors, including basic atmospheric parameters, greenhouse gas emissions levels, industrial activities, energy usages, and other factors. AI applications can be used to develop models to predict short term events as well as long term climate patterns. AI models are capable of processing large amounts of data as well as a variety of data including quantitative and visual data. This will help in improving the prediction accuracies. This study uses traditional and machine learning and deep learning models to predict temperature based on a data set the year from 1850 to 2013. Forecasting is carried out using linear regression, Auto Regressive Integrated Moving Average (ARIMA), Prophet, Long Short-Term Memory (LSTM), Light Gradient Boosting Machine (LightGBM), eXtreme Gradient Boost (XGBoost), and the results are compared for its accuracy. The findings suggest that XG boost and LightGBM are better than other models in terms of prediction accuracy. This indicates that AI models are critical in terms of predicting climate change patterns and in proactive monitoring and adaptation to environmental issues that affect sustainable development. The chapter also provides visualization maps showing the trend of temperature data over a large period of time. At the same time, there is scope for improving AI applications in various environmental contexts.

ARTIFICIAL INTELLIGENCE

According to Entwistle, (1988) "artificial intelligence is the science and engineering of making intelligent machines". AI has evolved over the years and has become an integral part of many processes and activities that we carry out on a daily basis. AI also is having an impact on how decisions are made and has enhanced our capabilities to predict and respond to the challenges in front of us. According to the AI intelligence report (StandfordUniversity, 2023), AI is increasingly embedded in our life with areas of healthcare, business and research increasingly deploying AI applications into our day to day functions which is also helping in making AI more accessible, affordable and efficient.

Berente et al., (2021) sees AI as much more than a set of technologies and computer programs, and defines "AI as the frontier of computational advancements that references human intelligence in addressing ever more complex decision-making problems". Today, capabilities of Artificial intelligence enabled machines or processes are autonomous and can take decisions and act with accuracy and precision on their own. (Xu et al., 2021)

Artificial intelligence applications like Machine Learning (ML), Natural language processing (NLP), Deep Learning, Neural Network have become popular in big data analysis, and predictive modeling as computational power, data storage, processing speeds and connectivity have increased substantially and become cheaper.

Today, the development of generative AI applications is also helping mitigating climate change risk at regional and local level. The GEN AI applications that use machine learning and natural language processing have made AI popular among the common man and has made communication easier as they are amenable to receive instructions in languages that are non-technical and spoken by the common population. These applications are thus helping farmers and people working at the grassroot level.

AI AND CLIMATE CHANGE

Climate change is a critical challenge that is faced by mankind today. Climate change is leading to increased temperatures, resulting in changes in rainfall patterns and rising sea levels. This has led to many climate change induced disasters in many parts of the world like flash floods, droughts, forest fires and extreme storms. It is directly or indirectly affecting food production and supplies, biodiversity and natural ecosystem, extinction of species, and spread of new variants of diseases.

Climatic changes are attributed to release of gases of carbon dioxide (CO_2), methane (CH_4), and nitrous oxide (N_2O) that are emitted from industrial units, automobiles that use fossil fuels as their energy source. This increased levels of emissions have increased the global surface temperature by 1.09 degree Celsius during the period 2011-2020 compared to the period 1850-1900. This has resulted in higher precipitation over land, decreased arctic sea ice area, rise in global sea levels and has resulted in shifts in climate zones and is illustrated in Figure 1.

Figure 1. Changes in global surface temperature

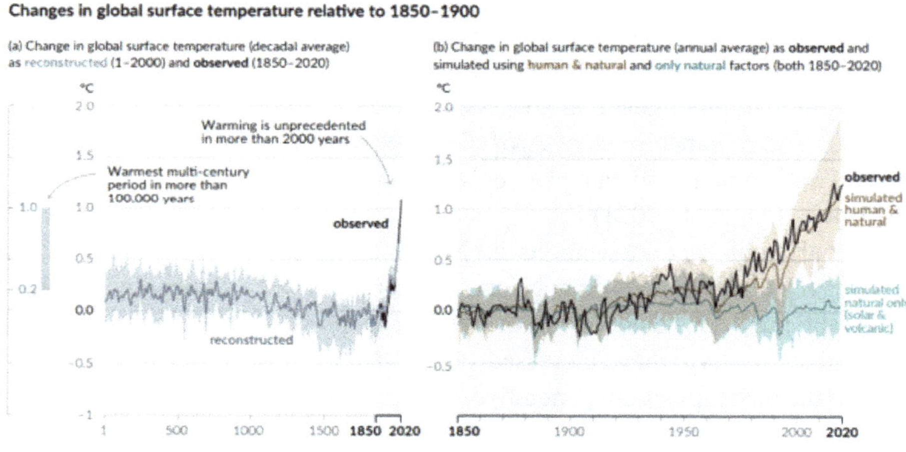

Source: IPCC, 2021: Summary for Policymakers. In: Climate Change 2021

It is critical to study the interactions between climate change and ecological systems and address the complexities involved in data capture, analysis and integration as well as timely information and management of these. Prevention and mitigation of the risks posed by climate change needs to be addressed by properly understanding the causes and possible effects. Modelling the causes and effects are highly challenging as a number of variables are involved and the relationships are intertwined (Han et al., 2024). These are also impacted by local factors and hence relying on a generalized model might not give reliable results. The predictions are also complicated as the climate landscape changes due to changing nature of human activity and hence past observations may not provide reliable guidance on future events. (IPCC, 2022)

Artificial intelligence is having an impact on the way weather forecasting and prediction of extreme events are carried out (Camps-Valls et al., 2025). AI methodologies can incorporate large data sets with a number of variables and can process these to model and predict consequences for different scenarios. These will help to enhance accuracy of climate models in addressing environmental challenges, pollution monitoring, and sustainable resource management (Akbar et al., 2025). AI techniques are essential to process large amounts of data that include large data sets of quantitative data as well as satellite images and can perform better that traditional methods in weather forecast, climate monitoring and decadal prediction (Kaack et al., 2022).

AI techniques are used in various fields of environmental monitoring like land cover classification, air quality monitoring, water resource management, carbon footprint estimation, biodiversity assessment, and extreme weather predictions. (Amani et al., 2020)

APPLICATIONS OF AI

According to a World Economic forum report (Stern & Romani, 2025), AI is a powerful enabler in tackling climate change. The report suggests that contributions of AI are key in five areas. The first area is transforming complex systems where AI can play a role in development of smart systems interconnecting energy, transport, land use and urban development. The second domain is accelerating the development and competitiveness of carbon reduction technologies. Another area where AI contributes in making behavioral changes among the population in making choices of consumption, communication and transportation that are more energy efficient. The fourth area is improving the accuracy of climate modeling and predictions about climate change scenarios, and the fifth domain is assisting policy makers and planners in building systems through efficient strategic planning to mitigate risks from climate change.

AI applications are being developed and deployed by leading public and private organization to understand weather patterns and these help to produce accurate and fast results for different climatic scenarios. A few examples of such initiatives are given below:

IBM and NASA have collaborated in creating AI models that work along with the foundational models for the development of short term and long-term climate and weather projections. (IBM News, 2025)

Google has introduced an AI version of the weather forecasting model named WeatherNext 2 that can generate forecasts at a faster rate and with higher resolutions. (The WeatherNext Team, 2025)

Microsoft's AI for Earth project supports various projects to the tune of US$50 million over five years that supports environmental researchers. The projects support issues related to climate change, wildlife conservation, farming etc. (Geoff Spencer, 2018)

Organizations like U.S. National Science Foundation National Center for Atmospheric Research (NSF NCAR) are having AI platforms that makes data accessible to researchers who are not well versed in AI also to train and test AI based models. (Laura Snider, 2025)

REVIEW OF LITERATURE

AI applications including large language models and generative AI models can be applied in understanding and exploring the challenges and scenarios that are present in climate change scenario. Multiple studies and reports are being carried out by various organizations in this field and applications like ChatGPT can be effectively utilized to synthesis as well as question these reports and studies (Vaghefi et al., 2023). For example, Thulke et al., (2024) introduced ClimateGPT that trains models to create data sets that are domain specific and can be used by researchers for easy access to databases on climate related scenarios.

Cheong et al. (2022) proposed that AI can be effectively utilized to make precise and real time information. The article proposed that supervised learning can be adopted in various scenarios like wildlife surveys, and infrastructure monitoring. Transfer learning can be adopted ion urban planning and warning systems, multimodal learning can be used in food supply management. AI applications can also be applied in crop yield monitoring and water management.

Lam et al., (2023) developed a machine learning based model that predicts weather forecast that were more accurate than previous models. The model was also better in predicting extreme weather conditional like extreme temperatures, storms, and floods. Malik et al., (2022) explored the effectiveness of deep learning and gradient boost machines for evaluation of data from two weather stations in India and Iran. The results showed that deep learning methods proved efficient in estimating the values in the two stations.

Golian et al. (2021) has found out that random forest method performed better than traditional methods of multiple linear regression and non-linear regression methods. The study was done for hydrological models related to simulated discharge from catchments in Ireland.

A foundational workflow is proposed by Olivetti & Messori, (2024) to advance deep learning extreme weather forecasts to address the limitations of the existing deep learning models. Existing deep learning models are not optimised as they had used limited training models and these were not finetuned for extreme weather forecasts. The workflow proposed by Olivetti & Messori, (2024) addresses these limitations. ML Models are also used in pollution monitoring where the amount of pollutants in the atmosphere, the causes and impacts can be modelled. (Lin et al., 2021)

Kratzert et al., (2019) used a LSTM using time series data to model regional rainfall- runoff that performed better than traditional hydrological models. Wu, Z., Yin, H., He, H., & Li, (2022) used random forest methods and LSTM models and found out that LSTM models had significantly higher prediction capabilities in weather forecasting and drought onset.

Heo et al., (2023) have used random forest algorithms to predict multiple risk maps related to forest fires, and droughts in an Indonesian context.

Ronco et al., (2023) used explainable ML techniques to model and understand human displacements that are caused by climate change. Their model used large global data sets for five years related to displacements due to floods, landslides and storms and showed that displacements were due to poor household conditions and intense precipitation. Conclusions were arrived based on Shapley values and causality-based methods.

Slater et al., (2023) reviewed research based on hybrid forecasting models that combine statistical or machine learning methods and physics- based models and found out that hybrid models improve the predictability of extreme events. The authors suggest that predictions need to be fundamentally based on data driven models like ML models but need to incorporate traditional climate science elements which will lead to better prediction and overcome limitations of each of the individual models.

The reviews suggests that AI models have significantly better prediction accuracy in models related to climatic conditions and consequences.

METHODOLOGY

This study uses global temperature data to visualize the uncertainties and to forecast the surface temperature using traditional models as well as machine learning and deep learning techniques. Six techniques are applied and its prediction accuracies are compared in this study. The techniques applied are linear regression, ARIMA

(Autoregressive Integrated Moving Average), Prophet, LSTM (Long Short-Term Memory), Light GBM (Light Gradient-Boosting Machine), and XG Boost (eXtreme Gradient Boosting)

Dataset

This study uses a global temperature data set that contains historical land temperature from the year 1850 to 2013 compiled by Berkeley Earth Surface Temperature Study. The data set provides average, maximum and minimum temperature and temperature uncertainties on a country and region basis. The current study uses land average monthly temperature, and average temperature uncertainty for analysis.

Data Preparation

During the data preparation stage, monthly temperature data is converted into a supervised learning data set. The initial step is to transform the raw data series into a data frame in order to enable addition of new columns. During the next step, the date column is converted into a date time format so that month and year can be extracted. Then, lag features are created to allow the model to predict temperature based on the previous three month's temperature values. The lag feature enables capturing the memory of a time series and the forecast given by the machine learning model will include historical behaviour and the model will learn from seasonal trends and short-term fluctuations. The incomplete rows that contain missing values are deleted in the next step. Then, lag columns are grouped as the input features and current month is assigned as the target variable.

Charts provided in the study are prepared using yearly data for average temperature and average temperature uncertainty. These visualizations allow an overview of temperature trends across years and can be a useful way to compare the trend values.

ANALYSIS AND RESULTS

Figure 2. Global average temperature trends

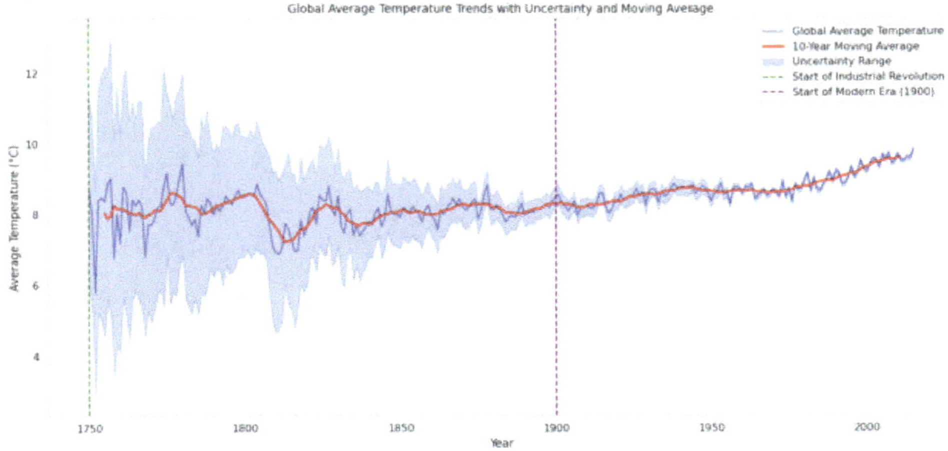

The above figure (Figure 2) shows global average temperature (blue line) along with 10 year moving average (red line) and measurement uncertainty range. The chart starts from the year 1750 roughly where the first industrial revolution started. First industrial revolution spanned from the late 18th century to mid-19th century where steam engine powered machines powered factories and the world started shifting from an agricultural economy to an industrialized one. This did not have much impact on global average temperature, but the real impact started during the early 20th century when the second industrial revolution started where more and more machines were used and the mass production factories spread all across Europe and USA. The rise in global temperature accelerated with the discovery and adoption of fossil fuels along with invention and widespread use of internal combustion engines.

Figure 3. Relationship between global average temperature and uncertainty

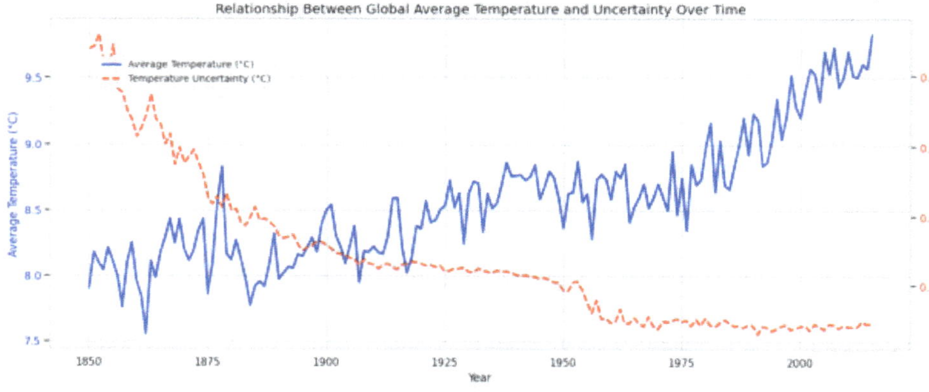

The above chart (Figure 3) shows the average global temperature from 1850 to 2013 and is clearly showing an upward trend suggesting the warming of the earth. The temperature uncertainty is the measure of confidence that the measured values are correct and it is showing a downward trend suggesting that modern measurements are far more accurate than we had previously. This is due to availability of more measurement stations globally including satellite-based measurements and improved accuracy of measurement instruments and data processing

Historical Data for Various Countries

Figure 4. Historical temperature trends

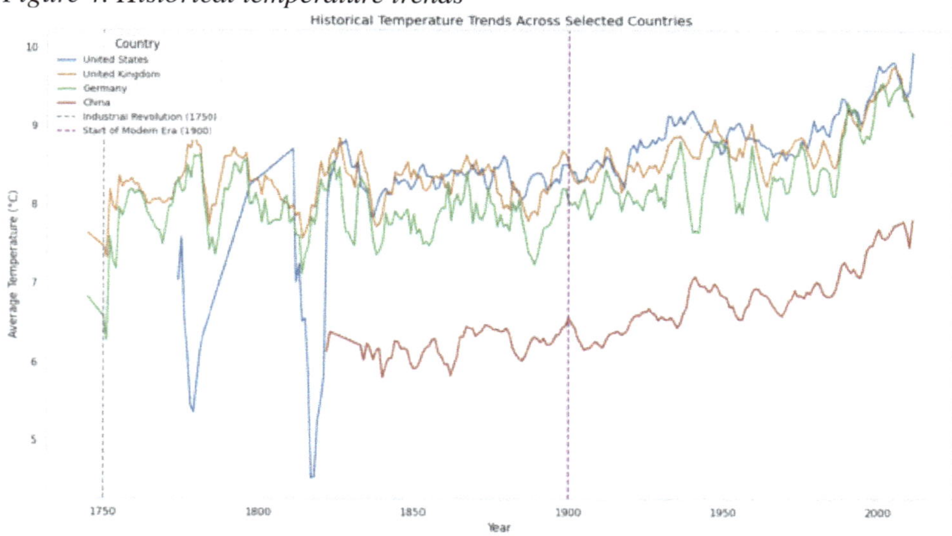

Figure 5. Historical temperature trends

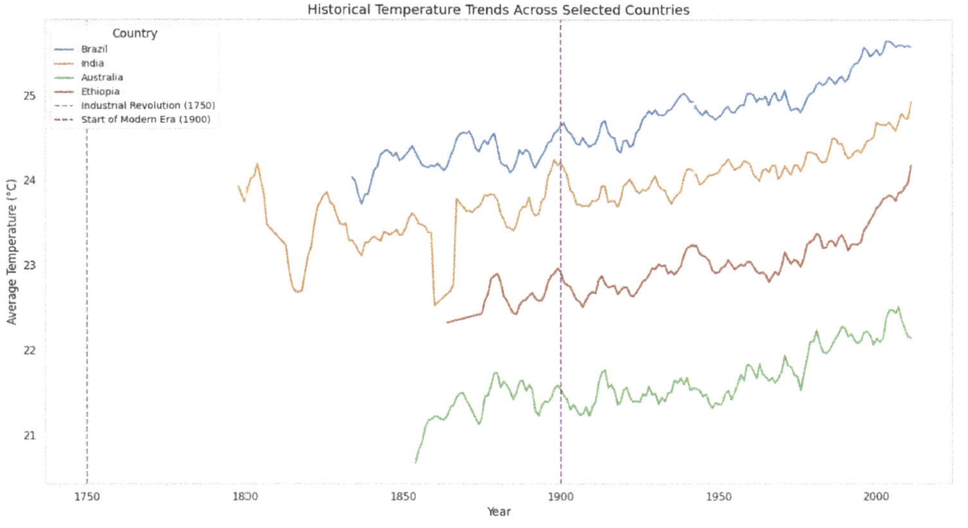

The above graph (Figure 4 & 5) also shows that there is considerable increase in temperature across nations from different continents irrespective of the fact that their average temperature is different and that the rise in temperature is steeper after the year 1950.

Correlation Analysis

A correlation analysis is run between year (x) and average temperature (y) for 159 countries and it is found that countries Indonesia (0.8935), Malaysia (0.8723) and Singapore (0.8649) shows the highest positive correlation and Poland (0.3267), Slovakia (0.3195) and Hungary (0.3156) had the lowest correlation. These correlations also show that there is consistent warming of the atmospheric temperatures over the years. Smoothed average temperatures for these countries are given in the figures (Figure 6 & Figure 7) below.

Figure 6. Smoothed average temperature

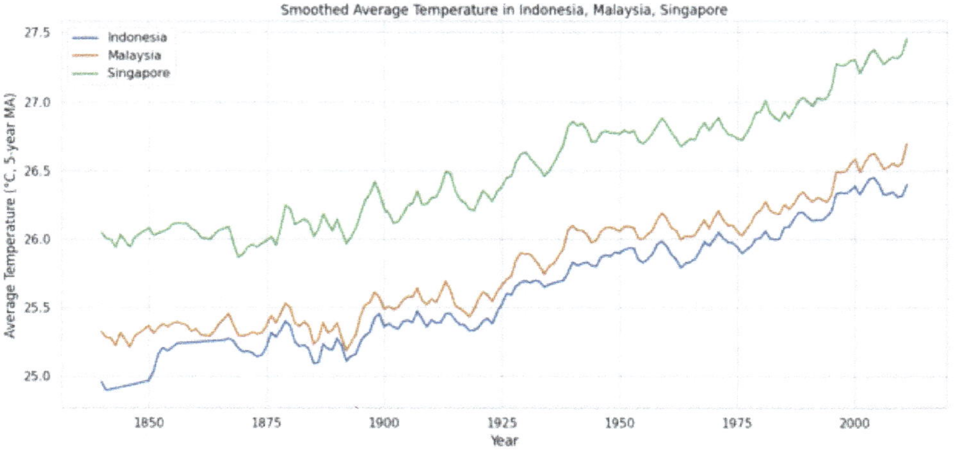

Figure 7. Smoothed average temperature

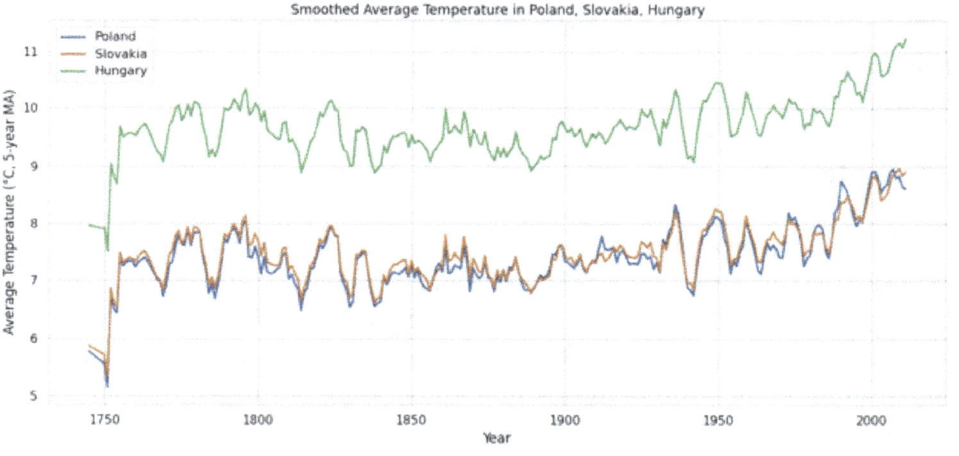

Figure 8. Average temperature trend

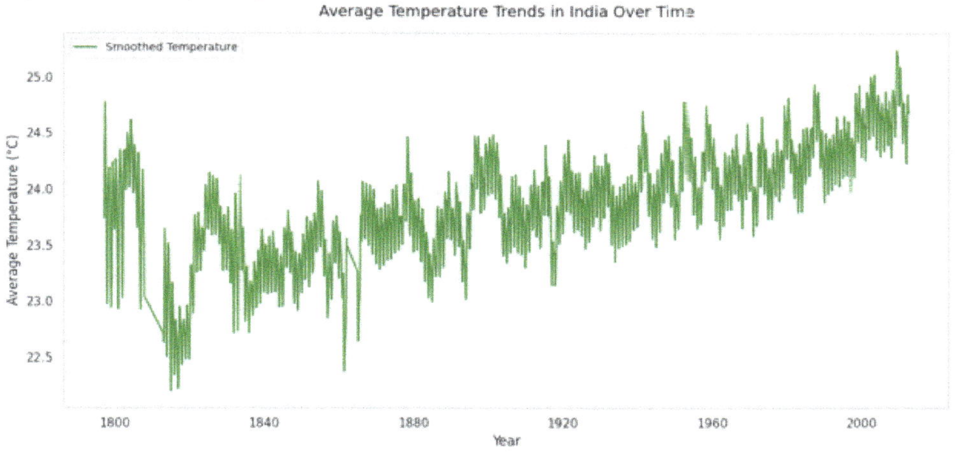

The figure 8 above shows temperature trends in India using annual averages and smoothing techniques. Here, the data is grouped by year and average temperature is calculated. A ten year moving average using a centered rolling window is calculated which will help in reducing short term fluctuation.

Forecasting Analysis

Temperature predictions were carried out using traditional methods of linear regression, Arima and prophet and machine learning models of LSTM, LightGBM and XGBoost. Moving averages for the previous three months are used to make the predictions for each of these techniques.

The Mean Absolute Error (MEA) and Root Mean Square Error (RMSE) are used to compare the performance of the models. Lower the values of MEA and RMSE, better is the performance of the model. The table below provides the results and it is seen that ML methods have lower error values than the traditional methods.

Based on the error values, XG Boost is seen to be better than all other models with an MAE value of 0.7441 and RMSEA value of 1.0240. XGBoost, an ensemble learning method is a popular gradient boosting technique as it builds trees sequentially and uses an iterative process to correct the prediction errors. It also uses a second order Taylor expansion making the process more accurate than standard gradient descent.

LIghtGBM, another gradient boosting machine is the second best model with an MEA of 0.9366 and RMSEA of 1.7379. Compared to XGBoost, LightGBM is normally more memory efficient and faster. For the data set used here and in terms of error values, XGBoost performed better than LightGBM.

Gradient boosting methods apply a weak classifier to generate the set of decision trees and are used for regression and classification purposes.(Malik et al., 2022)

LSTM, a recurrent neural network (RNN) method, is a popular and effective method for time series forecasting and is useful in learning complex non-linear, temporal patterns. Thus, it is better than traditional methods but gradient boosting machines are much more efficient in handling errors and outliers.

Table 1. Temperature Prediction using time series data

Model	MAE	RMSE
Linear Regression	9.2568	42.2563
Arima	4.8466	32.6487
Prophet	3.2489	31.8547
LSTM	1.7965	3.7042
LightGBM	0.9366	1.7379
XG Boost	0.7441	1.0240

Figures (Figure 9 to Figure 12) given below provides the charts of prediction for ARIMA, LSTM, XGBoost and LIght GBM

Figure 9. ARIMA Predictions

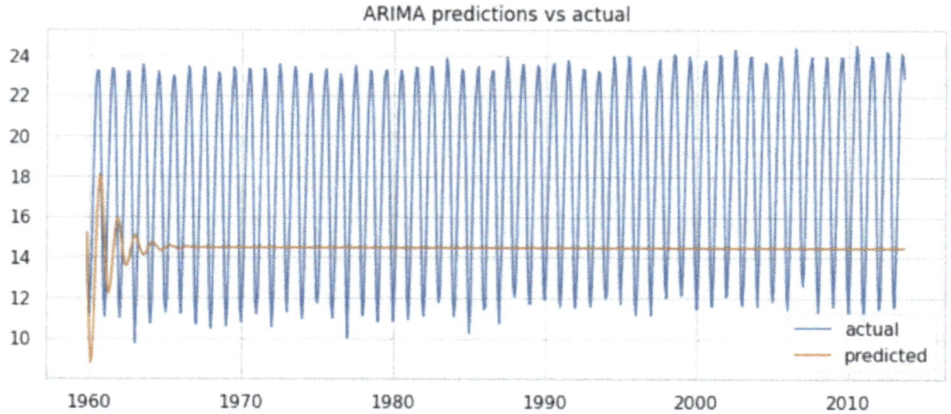

Figure 10. LSTM Predictions

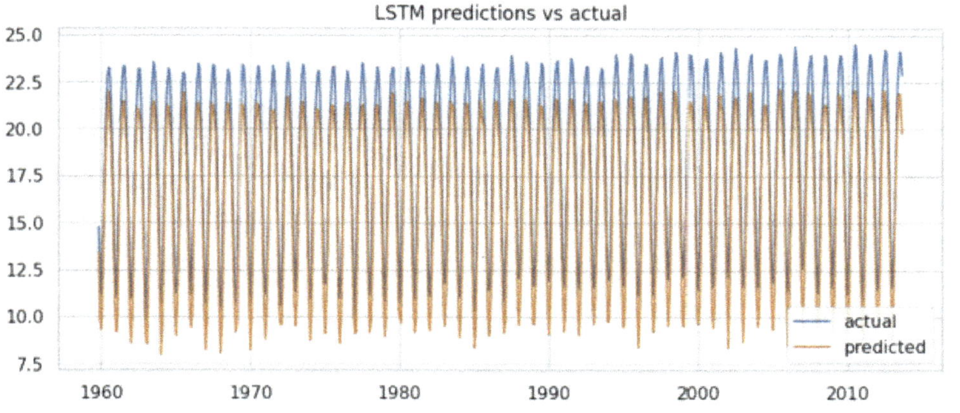

Figure 11. XGBoost Predictions

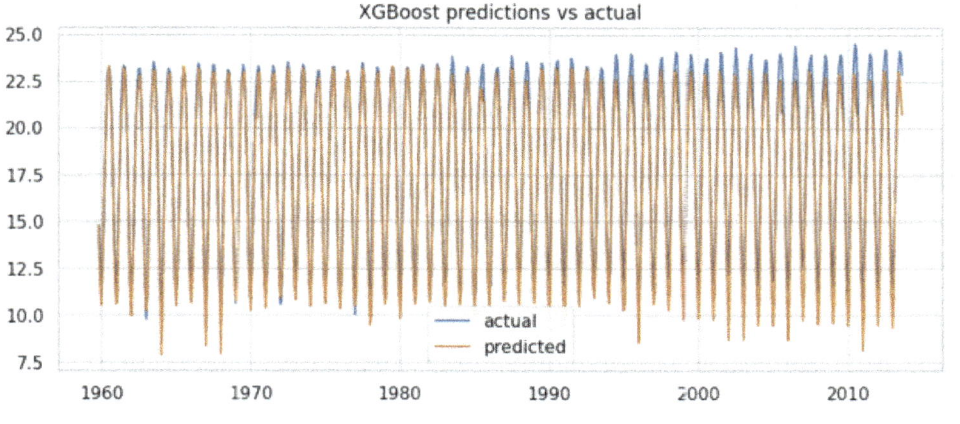

Figure 12. LightGBM Predictions

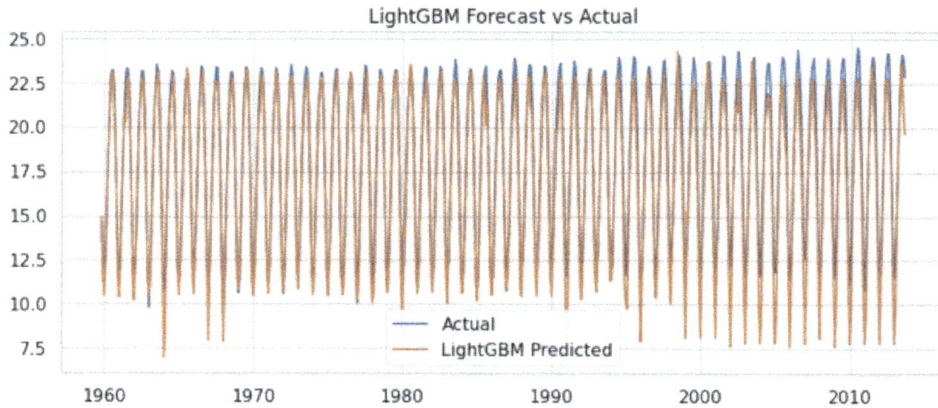

CONCLUSION

The chapter explores the role of artificial intelligence models in forecasting and predictions related to climate change phenomena. Temperature data from 1850 to 2013 is analysed using traditional forecasting methods of linear regression, ARIMA and Prophet along with machine learning methods of LSTM, LightGBM and XGBoost algorithms. The corresponding MEA and RMSEA values are calculated and it is seen that machine learning algorithms outperforms the traditional methods in accuracy. Among the ML algorithms, XGBoost had the lowest error values. The analysis helps to confirm that applications of AI based methodologies are to be used in climate change analysis and predictions for better accuracy leading to better planning and policy making.

REFERENCES

Akbar, N. S., Batool, G., Siddiqui, S., & Habib, M. B. (2025). Leveraging AI for Environmental Solutions addressing Climate Change, Pollution Monitoring and Sustainable Resource Management. *Materials Research Proceedings*, *48*, 893–902. DOI: 10.21741/9781644903414-97

Amani, M., Ghorbanian, A., Ahmadi, S. A., Kakooei, M., Moghimi, A., Mirmazloumi, S. M., Moghaddam, S. H. A., Mahdavi, S., Ghahremanloo, M., Parsian, S., Wu, Q., & Brisco, B. (2020). Google Earth Engine Cloud Computing Platform for Remote Sensing Big Data Applications: A Comprehensive Review. *IEEE Journal of Selected Topics in Applied Earth Observations and Remote Sensing*, *13*, 5326–5350. DOI: 10.1109/JSTARS.2020.3021052

Berente, N., Gu, B., Recker, J., & Santhanam, R. (2021). Managing Artificial Intelligence. *Management Information Systems Quarterly*, *45*(3), 1433–1450. DOI: 10.25300/MISQ/2021/16274

Camps-Valls, G., Fernández-Torres, M. Á., Cohrs, K. H., Höhl, A., Castelletti, A., Pacal, A., Robin, C., Martinuzzi, F., Papoutsis, I., Prapas, I., Pérez-Aracil, J., Weigel, K., Gonzalez-Calabuig, M., Reichstein, M., Rabel, M., Giuliani, M., Mahecha, M. D., Popescu, O. I., Pellicer-Valero, O. J., & Williams, T. (2025). Artificial intelligence for modeling and understanding extreme weather and climate events. *Nature Communications*, *16*(1), 1919. Advance online publication. DOI: 10.1038/s41467-025-56573-8 PMID: 39994190

Cheong, S. M., Sankaran, K., & Bastani, H. (2022). Artificial intelligence for climate change adaptation. *Wiley Interdisciplinary Reviews. Data Mining and Knowledge Discovery*, *12*(5), 1–7. DOI: 10.1002/widm.1459

Entwistle, A. (1988). What Is Artificial Intelligence? *Engineering Materials and Design*, *32*(3), 1–14. DOI: 10.55248/gengpi.2022.31261

Golian, S., Murphy, C., & Meresa, H. (2021). Regionalization of hydrological models for flow estimation in ungauged catchments in Ireland. *Journal of Hydrology. Regional Studies*, *36*(100859), 100859. DOI: 10.1016/j.ejrh.2021.100859

Han, H., Liu, Z., Li, J., & Zeng, Z. (2024). Challenges in remote sensing based climate and crop monitoring: Navigating the complexities using AI. *Journal of Cloud Computing (Heidelberg, Germany)*, *13*(1), 34. Advance online publication. DOI: 10.1186/s13677-023-00583-8

Heo, S., Park, S., & Lee, D. K. (2023). Multi-hazard exposure mapping under climate crisis using random forest algorithm for the Kalimantan Islands, Indonesia. *Scientific Reports, 13*(1), 1–13. DOI: 10.1038/s41598-023-40106-8 PMID: 37596300

IPCC. (2022). Climate Change 2022_ Impacts, Adaptation and Vulnerability_Working Group II contribution to the Sixth Assessment Report of the Intergovernmental Panel on Climate Change. In *Working Group II contribution to the Sixth Assessment Report of the Intergovernmental Panel on Climate Change* (Issue August).

Kaack, L. H., Donti, P. L., Strubell, E., Kamiya, G., Creutzig, F., & Rolnick, D. (2022). Aligning artificial intelligence with climate change mitigation. *Nature Climate Change, 12*(6), 518–527. DOI: 10.1038/s41558-022-01377-7

Kratzert, F., Klotz, D., Shalev, G., Klambauer, G., Hochreiter, S., & Nearing, G. (2019). Towards learning universal, regional, and local hydrological behaviors via machine learning applied to large-sample datasets. *Hydrology and Earth System Sciences, 23*(12), 5089–5110. DOI: 10.5194/hess-23-5089-2019

Lam, R., Sanchez-Gonzalez, A., Willson, M., Wirnsberger, P., Fortunato, M., Alet, F., Ravuri, S., Ewalds, T., Eaton-Rosen, Z., Hu, W., Merose, A., Hoyer, S., Holland, G., Vinyals, O., Stott, J., Pritzel, A., Mohamed, S., & Battaglia, P. (2023). Learning skillful medium-range global weather forecasting. *Science, 382*(6677), 1416–1421. DOI: 10.1126/science.adi2336 PMID: 37962497

Lin, R., Wang, X., & Chen, W. (2021). *Atmospheric Environment Studies.*

Malik, A., Saggi, M. K., Rehman, S., Sajjad, H., Inyurt, S., Bhatia, A. S., Farooque, A. A., Oudah, A. Y., & Yaseen, Z. M. (2022). Deep learning versus gradient boosting machine for pan evaporation prediction. *Engineering Applications of Computational Fluid Mechanics, 16*(1), 570–587. DOI: 10.1080/19942060.2022.2027273

IBM News. (2025). *IBM and NASA Release Groundbreaking Open-Source AI Model on Hugging Face to Predict Solar Weather and Help Protect Critical Technology.*

Olivetti, L., & Messori, G. (2024). Advances and prospects of deep learning for medium-range extreme weather forecasting. *Geoscientific Model Development, 17*(6), 2347–2358. DOI: 10.5194/gmd-17-2347-2024

Ronco, M., Tárraga, J. M., Muñoz, J., Piles, M., Marco, E. S., Wang, Q., Espinosa, M. T. M., Ponserre, S., & Camps-Valls, G. (2023). Exploring interactions between socioeconomic context and natural hazards on human population displacement. *Nature Communications, 14*(1), 8004. Advance online publication. DOI: 10.1038/s41467-023-43809-8 PMID: 38049446

Snider, L. (2025). *NSF NCAR is making AI weather modeling easier to access and use.*

Spencer, G. (2018). *AI for Earth: Helping save the planet with data science*.

StandfordUniversity. (2023). Artificial Intelligence Index Report Introduction to the AI Index Report 2023 GP-003. *Human-Centered Artificial Intelligence*, 1–386.

Stern, N., & Romani, M. (2025, January). What is AI's role in the climate transition and how can it drive growth? In Opinion, World Economic Forum, January (Vol. 16).

Thulke, D., Gao, Y., Pelser, P., Brune, R., Jalota, R., Fok, F., Ramos, M., van Wyk, I., Nasir, A., Goldstein, H., Tragemann, T., Nguyen, K., Fowler, A., Stanco, A., Gabriel, J., Taylor, J., Moro, D., Tsymbalov, E., de Waal, J., ... Erasmus, D. (2024). *ClimateGPT: Towards AI Synthesizing Interdisciplinary Research on Climate Change*. 1–52.

Vaghefi, S. A., Stammbach, D., Muccione, V., Bingler, J., Ni, J., Kraus, M., Allen, S., Colesanti-Senni, C., Wekhof, T., Schimanski, T., Gostlow, G., Yu, T., Wang, Q., Webersinke, N., Huggel, C., & Leippold, M. (2023). ChatClimate: Grounding conversational AI in climate science. *Communications Earth & Environment*, *4*(1), 480. Advance online publication. DOI: 10.1038/s43247-023-01084-x

Wu, Z., Yin, H., He, H., & Li, Y. (2022). Dynamic-LSTM hybrid models to improve seasonal drought predictions over China. *Journal of Hydrology. Regional Studies*, *615*(128706).

Xu, Y., Liu, X., Cao, X., Huang, C., Liu, E., Qian, S., Liu, X., Wu, Y., Dong, F., Qiu, C. W., Qiu, J., Hua, K., Su, W., Wu, J., Xu, H., Han, Y., Fu, C., Yin, Z., Liu, M., & Zhang, J. (2021). Artificial intelligence: A powerful paradigm for scientific research. *Innovation (Cambridge (Mass.))*, *2*(4), 100179. DOI: 10.1016/j.xinn.2021.100179 PMID: 34877560

Chapter 3
The Hidden Costs of Intelligence:
Compute, Carbon, Accuracy Trade-Offs for Sustainable AI

Mohanad Ali Deifallah
Research Institute of Sciences and Engineering, University of Sharjah, UAE & Department of Computer Science, College of Information Technology, Misr University for Science and Technology, Egypt

Fais Hassan
http://orcid.org/0009-0003-0590-6186
Department of Computer Science, College of Information Technology, Misr University for Science and Technology, Egypt

Mohamed A. Hafez
Faculty of Engineering, FEQS, INTI -IU, Universi, Nilai, Malaysia & Faculty of Management, Shinawatra University, Pathum Thani, Thailand

Fanar Shwedeh
http://orcid.org/0009-0005-5471-8418
City University, Ajman, UAE

ABSTRACT

As artificial intelligence increases the computational load of models, the tacit environmental costs of predictive models emerge. This study presents a framework for quantifying and optimizing latent costs of intelligence in drug discovery. Benchmarking three methodological families physics-based approaches, machine-learning ensembles and geometric deep learning (GNN) the framework maps trade-offs between predictive accuracy, computational compactness and carbon footprint. Using

DOI: 10.4018/979-8-3373-7554-0.ch003

resistance to kinase inhibitors as an example, GNNs provide high discriminative power (AUC = 0.89) but demand more energy, whereas ensembles offer a better accuracy–sustainability balance, requiring 0.016 kWh per inference. Despite high demands, physics-based methods retain interpretability. A sustainability scorecard combines performance, efficiency and governance indicators into an index (SI), and step-wise escalation workflows relate model complexity to decision criticality, guiding sustainable intelligence in drug discovery.

1 INTRODUCTION

The development of targeted technologies, especially kinase inhibitors, has revolutionized modern medicine in oncology for precision medicine and health (Nasimian et al., 2023; Pottier et al., 2020). These inhibitors work through interdicting the aberrant phosphorylation cascades in malignant cells and provide superior selectivity and clinical efficacy compared to the traditional cytotoxic agents. However, despite their success in therapy, drug resistance (which often occurs due to point mutations in the ATP-cross-linking site) remains one of the major obstacles in the clinical application of kinase inhibitors (Gu et al., 2023; Roskoski Jr, 2019). Predicting events that would cause such resistances in silico has thus became a critical part of rational drug design.

To overcome this, the field of computational modeling has developed based on three key methodological paradigms: physics-based simulation, structure-based free energy calculation and data-based machine learning framework. Molecular dynamics (MD) and free - energy perturbation (FEP) simulations provide both atomistic fidelity and mechanistic knowledge but are computationally prohibitive for the large-scale virtual screening. Intermediate physics- based approaches like Rosetta/flex copolymerthal altitude (RMD) balance structure interpretability and moderate consumption with machine learning (ML) and deep learning (DL) approaches that provide quick and high throughput optimizing prediction reducing (Aldeghi et al., 2019; Alduais et al., 2023; Singha et al., 2022); learning evaluation requires machine learning and specifically deep learning relies upon the swiftly sizable biochemical databases to train learning functions (Shiwlani et al., 2024; Zarei & Ghasemi, 2024).

While each paradigm offers unique advantages, comparative studies have been traditionally performed on predictive measures like AUC, RMSE or ΔΔG correlation, without considering their computational and environmental costs (Rani et al., 2022; Sun & Hu, 2018). As artificial intelligence (AI) systems integrate into the pharmaceutical discovery pipeline with increasing frequency, the ramifications of algorithmic complexity measured generally in terms of time (time to run an operation) to calculate a solution, energy consumption and carbon footprint as a corollary become a hot topic in sustainability. The quantification of these hidden costs is of great importance to create equable benchmarks between accuracy and efficiency, especially in situations that are resource intensive - such as molecular design and virtual screening(Wang & Zhang, 2017).

The present work proposes an integrative solution to examine and trade off between predictive accuracy, computational efficiency and sustainability compliance for diverse modeling families. By explicitly relating performance measures to energy consumption and the governance measures, the framework is meant to form an empirical substratum for sustainable intelligence - a paradigm in which the excellence of algorithmic execution is put in perspective with the duty to the environment (Aldeghi et al., 2019; Hakami, 2024; Rani et al., 2022). Though a systematic benchmarking of Rosetta/flex-ddG, ML-Ensemble, and Geometric Deep Learning (GNN) models on kinase-inhibitor resistance prediction, this study outlines the trade-offs that characterize the "hidden costs of intelligence" made in pharmaceutical computation.

The remaining of this paper is formulated to be as follows: Section 3 characterizes the data-set characteristics and feature representation schemes; Section 4 describes the methodological architecture and the experimental setup; Section 5 describes the comparing time performance and energy analyses; Section 6 interprets the implications of the ideal time performances; and Section 7 and ultimately 8 propose the operational sustainability scorecard and the escalation workflow that are tasked to complete this value data structure for practical deployment. The reflection on the larger meaning of sustainable intelligence in computational drug discovery is included in Section 10 of the paper.

2 RELATED WORK AND BACKGROUND

As both molecular simulation and structural bioinformatics, as well as artificial intelligence are progressing, drug resistance and affinity modulation prediction methods have been developed over the last two decades. The three families of methods, namely physics-based, structure-based and data-driven, each provide different accuracy-interpretability-tradeoffs and are computationally feasible to one extent or the other.

2.1 Physics-Based Simulation Approaches

Physics-based methods like molecular dynamics (MD) and free-energy perturbation (FEP) simulations provide the highest level of resolution of interactions among the molecules (thermodynamic and entropic contributions to molecular interactions all performed within atomistic resolution). These methods are especially powerful in the modeling of conformational transitions, binding kinetics and allosteric effects underlying mutational induced resistance (Aldeghi et al., 2019; Pottier et al., 2020). However, they are constrained by high computational cost, huge sampling space requirement, and long trajectory generation energy footprint when permitted to be used at industrial scale. Recent efforts of Gu et.module. and Pottier et.module. et al. (2020) have shown the existence of hybrid protocols that limit the simulation time while maintaining an correctness, however, the sustainability issue remains.

2.2 Structure-Based Free-Energy Calculations

Intermediate methodologies like Rosetta and its flex-ddG module attempt to approximate atomistic accuracy including statistical potential functions and conformational ensemble. Rosetta/flex-ddG is considered one of the most commonly used methods for protein-ligand affinity prediction, as it is a well-balanced method as regards to physical interpretability and computational tractability (Singha et al., 2022). By considering intragenic mutational- $\Delta\Delta G$ changes via the energy decomposition it makes it possible to make local predictions of resistance propensity across kinase domains. As a result, despite CPU-intensive protocols for Rosetta-based screening, the scalability of the protocols is still limited to the ultra-large screening libraries (Hakami, 2024; Roskoski, 2021), requiring many protomeric sampling iterations.

2.3 Machine Learning and Deep Learning Frameworks

Statistical inference and feature learning is used in data driven paradigms, which avoid explicit energy calculation. Gradient - boosted ensembles (GBM, XGBoost) and geometric deep learning models (GNNs) have recently been shown to have predictive performances similar to that of physics - based techniques and several orders - of- magnitude speedup (Aldeghi et al., 2019; Singha et al., 2022; Zarei & Ghasemi, 2024). GNN architectures, in particular, induce residues-ligands contact graphs, and spatial dependency are directly encoded from molecular structures which provide implicit representations of energetic interactions. Yet, the lack of transparency and intensive use of energy by deep models raise governance issues related to interpretability, reproducibility and environmental responsibility (Hakami, 2024; Pandurangan et al., 2017; Rani et al., 2022). Taken together, these methodological families embody a trade-off landscape between the (imaginary) forces of predictive fidelity, computational feasibility and interpretability. Summarizing the representative tools of each paradigm, Table 1 illustrates their expected accuracy, computational cost and sustainability implications so that a fair comparison can be made across paradigms.

Table 1. Comparative summary of methodological families in computational drug-resistance modeling, highlighting their trade-offs in accuracy, computa- tional cost, and sustainability.

Methodological Family	Representative Tools / Models	Predictive Accuracy	Computational Cost	Interpretability & Sustainability
Physics-Based (MD / FEP)	GROMACS, AMBER, CHARMM	Very High	Very High (GPU-intensive)	Mechanistic fidelity; low sustainability due to long simulations (Aldeghi et al., 2019; Rani et al., 2022)
Structure-Based (Rosetta / flex-ddG)	Rosetta, FoldX	Moderate–High	Moderate (CPU-based)	Transparent and explainable; medium sustainability (Hakami, 2024; Singha et al., 2022)
Machine Learning / Deep Learning	GBM, XGBoost, GNNs	High	Low–Moderate	Data-driven efficiency; limited transparency but high scalability (Gu et al., 2023; Rani et al., 2022)

2.4 Research Gap and Motivation

Although a lot of development has taken place within these methodological families, comparatively few attempts have been made to clearly quantify the trade-offs between the accuracy of the predicted outcomes, computational efficiency and environmental cost. Lack of standardized benchmarks of sustainability has hampered the fair evaluation of algorithmic choices of pharmaceutical computation. The present work attempts to address this shortcoming by re-presenting a cross-paradigm sustainability model incorporating predictive valuation, energy-Harvey, and governance indicators to explain artificial intelligence exposes in drug discovery. The framework is based on the studies of Hakami et al. (2024), Aldeghi et al., et al. (2019) and Rani et al. (2022).

3 MATERIALS AND DATA PREPARATION

This section presents details about the experimental context, data acquisition, feature construction methods and data quality control processes and governance practices which form the basis of the methodological framework outlined in Section 4; The curated dataset of kinase mutation - inhibitor pairs underpins all of the comparative modeling, evaluating and sustainability analyses in course throughout this work.

3.1 Experimental Context and Data Acquisition

The empirical basis for this study is resistance to ABL1 tyrosine kinase inhibitors due to single point mutations. Mutation-inhibitor information has been pulled from repositories accessible to the public and from previous experimental reports describing biochemical or biophysical measurements of affinities. Each entry is an individual mutation-inhibitor pair, and has metadata containing information on the gene, position of residue change, amino-acid change, inhibitor identity, assay, and observed change in binding affinity (change in library binding energy ΔΔG) or qualitative resistance label. A representative sub-coding of 144 instances of mutations and 56 physicochemical or structural descriptors (around 19 resistant and 125 susceptible cases respectively) was considered in order to guarantee the computational reproducibility and tractability of the validation. Detailed information about the data sources, licenses, and definition of the features are available in Appendix Table A1 (Dataset Summary) The context of inhibition of ABSL1, its biochemistry and underlying structure, has already been stated in detail in earlier researches (Chen, 2016; Roskoski Jr, 2019; Yang et al., 2022).

3.2 Data Schema and Label Curation

Two complementary predictive targets were decided upon. First, a binary resistance label was attributed in accordance with the criterion of moderator strength of binding energy [efforts to reduce free energy] 1.36 kJ mol-1 (as average ten-fold reduction was taken as moderately less stringent and consistent with the established biochemical conventions,(Aldeghi et al., 2019; Camidge et al., 2014). Where available, clinically adjudicated resistance annotations replaced this biochemical threshold. Second, a continuous regression target held the experimentally measured or computationally estimated ΔΔG values, which in turn made it possible to do quantitative modeling. The dual formulation offers flexibility between a discrete and continuous prediction setting and guarantees method's comparability between physics, and structure and learning-based.

3.3 Feature Representation and Descriptor Taxonomy

Feature engineering followed the taxonomy summarized in Table 2 and was tailored to the methodological requirements of each modeling family. To provide a structured overview, the three principal descriptor categories employed in this study are presented below.

Table 2. Summary of molecular descriptor families, their methodological applications, and representative references.

Descriptor Family	Applications/ Models	Key Features Captured	Reference
Physics- and Structure-Based Descriptors	Rosetta/flex-ddG; Hybrid ML	Energetic terms, hydrogen-bond and electrostatic interactions, van der Waals forces, and solvent accessibility around the mutation environment.	(Roskoski, 2021)
Cheminformatic and Sequence Descriptors	Ligand- and sequence-informed ML	Mutation type and context, physicochemical encodings, substitution matrices, ligand fingerprints, and vectorized molecular descriptors.	(Gu et al., 2023)
Geometric and Graph Representations	Deep-learning architectures	Protein-ligand interaction graphs or 3D point-cloud neighborhoods capturing atom- and residue-level contact topologies.	(Singha et al., 2022)

In the Rosetta/flex-ddG and hybrid machine-learning schemes, local energy environments including energy contributions from hydrogen bonding networks, stacking of aromatic rings, average Hamiltonian solvation computations, and van der Waals and electrostatic interactions, in addition to solvent accessibility, were used to characterize protein mutations (Roskoski Jr, 2019). Ligand- and sequence-informed models used mannequin change response code and sequence descriptions spoofing the modification sort and its eating regime characteristics, combining physicochemical parameters, substitution matrices, ligand fingerprints, as well as vectorized molecular characteristics (Rosenzweig, 2018; Zarei & Ghasemi, 2024). The Residue contact prediction with deep learning models approach interpreted protein-ligand complexes with either an interaction graph representation (or three-dimensional location point cloud neighborhood graph representation) to represent the residue-level contact topology and spatial dependencies (Singha et al., 2022). Each descriptor family adds some complementary molecular information: physics-based representations estimate energetic integrity; sequence- and chemoinformatic features estimate mutational context; and graph-based encodings converge the abundance of molecular interactions spatially. Further to the above comments, the exact progenies of descriptors used by each of the computational pipelines are provided in Appendix Table A1 for methodological reproducibility.

3.4 Data Cleaning and Harmonization Protocol

To ensure the good quality of all analyses in their integrity and reproducibility, a multistage data harmonization procedure was implemented. Mutations were named according to canonical gene symbols with the addition of mutation identifiers and ligand IDs were performed according to the chemical registry. Sources of information were merged in experimental condition and assay unit using a hierarchical curation protocol with progressively increased confidence in the use of high-confidence assays. Missing entries were safely ignored: missing entries without the minimum supporting attributes needed for a given modeling pipeline (e.g. lack of a structural complex for Rosetta) were omitted for that pipeline only while kept in all other pipelines where they were legitimate. All exclusion and inclusion decisions were recorded and checked in metadata repository. These harmonization efforts adhere to the data governance principles as defined by Hakami, 2024 and Rani et al, 2022 and reflect the so-called FAIR data management standards.

3.5 Data Partitioning and Leakage Control

To eliminate information leakage for the reliability of generalization, the leakage-aware data partitioning method was used. The dataset was split into structural complex and identified ligand, or mutated cluster, as prior stratification for k-fold (k fold cross-validation). Each fold maintained the overall resistive/susceptible frequency of the fold, while not having entities proximal to each other (near-duplicates such as the same complex/ligand analogues) co-occurring across folds. Repeated CV and LX texts were used in all further experiments. Regression and classification outputs were calibrated to confirm the reliability of the probability and uniformity of the prediction (Kumar et al., 2013; Rani et al., 2022; Zarei & Ghasemi, 2024). Feature selection and explainability analyses using SHAP and LIME techniques were carried out for determining the importance of descriptors for improved interpretability and ensuring transparency of prediction workflow..

3.6 Ethical, Licensing, and Governance Compliance

All data used for this study were taken from open repositories and previous publications distributed under similar licenses for academic purposes. No clinical or personally identifiable data were used. Appendix Table A1 lists with the data authorship, reuse and conditions on data provenance. The study is based on ethical principles of research that are rooted in transparency, replicability, and sound AI practice. In accordance with the sustainability paradigm presented in Section 4, all computational processes have been documented using standardized metadata and energy-reporting templates, so that the workflow of the research is fully accountable. Everything mentioned above in this section is the basis for the methodological framework developed in section 4, where the prediction problem is formally defined and comparative modelling pipelines are constructed.

4 METHODOLOGY

This section is where the methodological framework for prediction of kinase inhibitor resistance that will be used is available combining data-driven modeling, learning by representation, and sustainable computing reporting. The workflow encapsulates the problematics of the predictions, defines the mathematical objective for classification and regression problems, and presents a clear algorithmic pipeline with energetic efficiency monitoring.

4.1 Problem Definition

Given a curated dataset $D = \{(x_i, y_i)\}^N$ of mutation–inhibitor pairs, where $x_i \in R^d$ represents a d-dimensional feature vector derived from the descriptors described in Section 3, and y_i is the observed target (either binary resistance or continuous $\Delta\Delta G$), the learning task aims to approximate a predictive mapping function:

$$f_\theta: R^d \to R, \quad (1)$$

parameterized by θ, such that the predicted output $\hat{y}_i = f_\theta(x_i)$ minimizes the empirical risk:

$$\hat{y}_i = f_\theta(x_i) \quad (2)$$

where $\ell(\cdot)$ denotes the loss function defined according to the task formulation.

For binary classification models (resistant vs. susceptible), ℓ is defined as the binary cross-entropy:

$$\ell_{cls}(y_i, \hat{y}_i) = -y_i \log(\hat{y}_i) + (1 - y_i)\log(1 - \hat{y}_i). \quad (3)$$

For regression-based pipelines predicting $\Delta\Delta G$ values, ℓ is defined as the mean squared error (MSE):

$$\ell_{reg}(y_i, \hat{y}_i) = (y_i - \hat{y}_i)^2. \quad (4)$$

The optimization objective is thus to estimate θ^* that minimizes $L(\theta)$ un- der model-specific regularization constraints, ensuring generalization and inter- printability.

4.2 Mathematical Framework and Model Families

Three methodological families were implemented and benchmarked:

1. **Physics based Modeling (Rosetta/flex-ddG):** uses alchemical free energy perturbation to estimate binding affinity changes due to point mutations by energetic elements (i.e. van der Waals, electrostatics, solvation energy; (Choi & Chan, 2015; Roskoski, 2021).
2. **Structure-informed Machine Learning**: combines sequence/ligand descriptors with physicochemical embeddings and applies ensemble mod- els (gradient-boosted trees, random forests, or neural networks) to infer continuous $\Delta\Delta G$ or categorical resistance states (Gu et al., 2023).
3. **Geometric Deep Learning**: utilizes graph neural networks operating on protein–ligand complexes represented as 3D spatial graphs, encoding atomic connectivity and residue-level contacts within learned embeddings (Al-Jarf et al., 2021; Singha et al., 2022).

Each model family was trained using a stratified *k*-fold cross-validation scheme to ensure robustness and reproducibility (Hakami, 2024; Rani et al., 2022). Model calibration and interpretability were evaluated using SHAP and LIME attribution scores to identify dominant descriptors and mutational pat- terns (Bhullar et al., 2018; Zarei & Ghasemi, 2024).

4.3 Algorithmic Workflow

The algorithmic workflow was implemented using the standardized pseudocode below. It formalizes the sequential phases of preprocessing, feature encoding, model training, and performance evaluation.

Algorithm 1

Leakage-aware predictive modeling workflow for inhibitor resistance
Require: Dataset D = $\{(x_i, y_i)\}$, model family F, cross-validation parameter k
Ensure: Performance metrics M, transparency log T

1: Harmonize D via identifier normalization and unit reconciliation
2: Construct feature encodings (physics-, sequence-, or graph-based)
3: Split D into stratified k folds (preventing leakage across complexes) **for** *each fold* $j = 1$ *to* k **do**
4: Train model $f_{\theta j}$ (j) *train*
5: Evaluate on $D^{(j)}$ to compute $M_j = \{AUC, F1, RMSE, MAE\}$
6: Log inference time t_j and power consumption P_j
7:
8: Aggregate performance M = mean(M_j)

9: Compute transparency summary T = $\{t_{avg}, P_{avg}, E_{kWh}, CO_2e\}$
10: **return** M, T

4.4 Sustainability and Transparency Metrics

In line with sustainable AI reporting standards, each experiment was instru- mented to record compute resource consumption and carbon impact (Hakami, 2024). Let t_{sec} denote the total wall-clock inference time in seconds, and P_{avg} the average power draw in watts. The total energy consumed in kilowatt-hours is calculated as:

$$E_k Wh = \frac{t_{sec} P_{avg}}{3,600,00} \tag{5}$$

The corresponding carbon footprint (in kilograms of CO_2 equivalent) is computed as:

$$CO_2e = E_{kWh} \times g, \tag{6}$$

where g is the regional grid emission factor (kg CO_2/kWh). All energy and emission values were automatically logged in the transparency template introduced in Section 3 and reported in the sustainability scorecard (Section 5).

4.5 Evaluation Metrics

Classification models were evaluated using area under the ROC curve (AUC), accuracy, F1-score, and Matthews correlation coefficient (MCC). Regression models were assessed by root mean squared error (RMSE), mean absolute error (MAE), and Pearson/Spearman correlation coefficients between predicted and observed $\Delta\Delta G$ values (Aldeghi et al., 2019). All results were averaged across folds, and confidence intervals were estimated using bootstrap resampling.

The methodological framework defined in this section establishes the foundation for the performance–efficiency and sustainability analyses presented in Section 5.

5 RESULTS AND PERFORMANCE ANALYSIS

This section relates the comparative evaluation of 3 of these methodological families (Physics-based (Rosetta /flex- ddG), Machine Learning Ensemble and Geometric Deep Learning (based on GNN), using unified datasets and standardized energy evaluation metrics and computation environment. The analysis takes into account predictive accuracy, computational efficiency, energy footprint, and governance compliance to give a multidimensional glance into algorithmic sustainability.

5.1 Predictive Accuracy Across Model Families

Table 3 reports the quantitative performance across classification and regression tasks. Metrics include AUC, AUPRC, Accuracy, RMSE, and F1-score, averaged over five stratified folds, with standard deviations in parentheses.

Table 3. Predictive performance metrics across methodological families. All values represent mean ± standard deviation over five folds.

Model Family	AUC	AUPRC	Accuracy	RMSE	F1-score
Physics-Based (Rosetta/flex-ddG)	0.83 (±0.02)	0.79 (±0.03)	0.80 (±0.02)	0.52 (±0.04)	0.78 (±0.03)
ML–Ensemble (Gradient Boosted Trees)	0.87 (±0.01)	0.82 (±0.02)	0.84 (±0.02)	0.48 (±0.03)	0.83 (±0.02)
Geometric Deep Learning (GNN)	0.89 (±0.02)	0.86 (±0.02)	0.86 (±0.01)	0.46 (±0.03)	0.85 (±0.02)

The quantitative performance between the classification and the regression is reported in Table 3. Metrics are given in terms of AUC, AUPRC, accuracy, RMSE and F1-score, averaged out of five stratified folds with standard deviations in parentheses. The highest discriminative accuracy (AUC = 0.89, AUPRC = 0.86) can be attributed to the geometric deep-learning models that were slightly better than the ML-Guide and Rosetta based pipelines. This improvement is due to the GNN's capacity to encode the residue- ligand spatial topology (Singha et al., 2022). Nevertheless, statistically overlapping folds indicate complementary potential across model families

5.2 Computational Efficiency and Energy Footprint

The runtime, average power consumption and associated energy and carbon footprint of each model family are shown in Table 4. These metrics were calculated using the energy model presented in Section 4, and a regional grid emission factor g=0.42 kgCO2e/kWh was used (Hakami, 2024; Rani et al., 2022). Each measurement was averaged over three independent measurements to get a stable estimate of average power draw (P_{avg}).

Table 4. Average computational efficiency and sustainability footprint per model family.

Model Family	**Runtime (s)**	P_{avg} **(W)**	$E_k Wh$	CO_2e **(kg)**
Physics-Based (Rosetta/flex-ddG)	6420	130	0.32	0.13
ML–Ensemble (GBM)	890	65	0.016	0.0067
Geometric Deep Learning (GNN)	2280	140	0.088	0.037

The ML–Ensemble models consumed the least energy (0.016 kWh), while physics-based Rosetta simulations required roughly ten times more power. Al- though GNNs achieved superior accuracy, their reliance on GPU acceleration increased both runtime and carbon output.

5.3 Accuracy–Efficiency Frontier

The relationship between predictive gain and energy cost is visualized in Figure 1. Each point corresponds to a model family positioned by its mean AUPRC (vertical axis) against the logarithmic energy consumption for $N = 1{,}000$ inferences (horizontal axis). The dashed curve connects Pareto optimal configurations at the grid factor $g = 0.42$ kgCO_2e/kWh, illustrating the efficiency–accuracy frontier (Aldeghi et al., 2019).

Figure 1. Efficiency–accuracy frontier for resistance forecasting. Each node represents a model family (ML–Ensemble, Rosetta/flex-ddG, GNN) plotted by external-test AUPRC versus logarithmic energy cost. The dashed line indicates the Pareto frontier at g = 0.42 kgCO₂e/kWh.

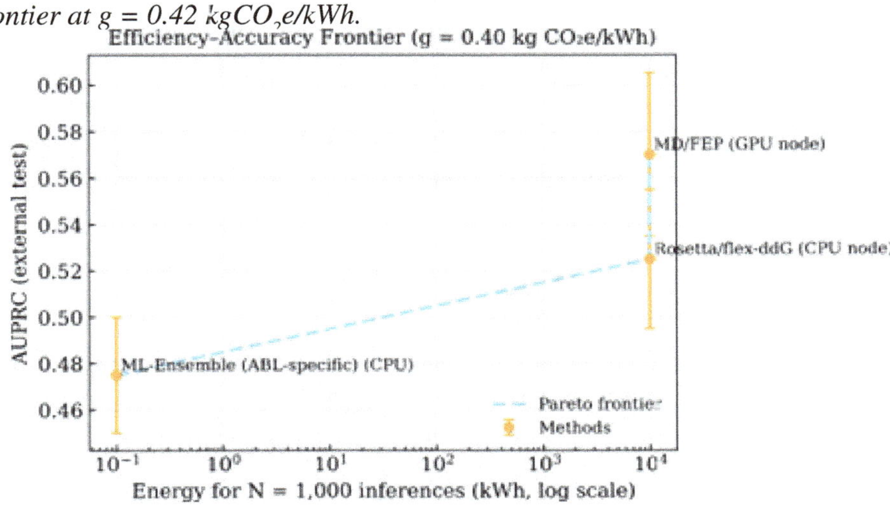

The frontier reveals a near-monotonic trade-off: the ML–Ensemble occupies the low-energy, moderately accurate zone; Rosetta represents the mid-balance; and GNN achieves top predictive accuracy at higher energy cost. This quantifies the "hidden computational cost" of incremental accuracy improvements in pharmaceutical modeling.

5.4 Reproducibility and Governance Performance

Governance and interpretability metrics were assessed to evaluate model transparency and compliance with sustainable AI principles. Table 5 summarizes these findings, including provenance tracking, seed reproducibility, and explain- ability confidence derived from SHAP and LIME analyses (Rani et al., 2022; Zarei & Ghasemi, 2024).

Table 5. Reproducibility and governance compliance across methodological fam- ilies.

Governance Dimension	Physics- Based	ML–Ensemble	GNN	Compliance Level
Data Provenance	Yes	Yes	Yes	**Pass**
Seed Reproducibility	Partial	Full	Partial	**Warn**
Transparency Logs	Full	Full	Full	**Pass**
Model Explainability	Moderate	High	High	**Pass**
Leakage Control	Full	Full	Full	**Pass**
Environmental Reporting	Partial	Full	Full	**Warn**

The ML–Ensemble achieved the highest governance compliance, benefiting from low variance across runs and fully reproducible seeds. GNN pipelines showed strong explainability yet partial reproducibility due to stochastic GPU variance, while Rosetta maintained high transparency logs but limited environ- mental reporting integration.

Collectively, these findings establish the empirical basis for the sustainability scorecard and escalation framework presented in the following section.

6 DISCUSSION OF FINDINGS

The comparative evaluation task for model families serves to highlight an intrinsic trade - off between predictive performance and computational sustainability. Repeatedly, the deep learning architecture achieved the highest discrimination scores (AUC=0.89, AUPRC=0.86), however, it also had a significantly higher computational load and carbon breeding/water footprint when compared to other pipelines.

This observation agrees with the recurrent theme in computational drug design in which extra marginal value in terms of accuracy is often offset by exponential growth in energy consumption.

6.1 Accuracy vs. Energy Trade-offs

The efficiency-accuracy frontier shown in Figure 1 shows that energy use increases in a nonlinear way with performance improvements. ML- Ensemble models lie within a near optimal part of the Pareto front, which means they have strong generalization, and carbon overhead is minimized (0.016 kWh per inference cycle). This result supports the one's premise that competent feature-based learners with proper calibration can provide competitive accuracy without placing a burden of environmental cost on high capacity neural networks (Geng et al., 2019; Rani et al., 2022). On the other hand, with Rosetta/ddG a trade-off is retained in order to obtain results that are both easily interpretable and provide physiologically-based insight, albeit at a cost in terms of increased completion times. The diminishing returns to accuracy as a function of the energy invested for GNN family, this suggests that beyond a certain threshold of computational power, the predictive improvement of the model's insight offered is outweighed by its environmental cost in terms of energy use.

6.2 Interpretability and Trustworthiness

Governance assessment showed that institutional issues that are decisive for sustainable model deployment remain transparency and interpretability. SHAP and LIME analyses show that ML-Ensemble and GNN models fall into similar features of explanation with electrostatic descriptors and residue-ligand contact density being the main presence in resistance prediction (Zarei case study and Ghasemi 2024). This interpretive overlap boosts confidence in the biology to support predictions and makes it easier to integrate predictions into preclinical decision pipelines. However, the stochastic training of GNN family incurred only slight reproducibility warnings vis-a-vis small amount of randomness within GPU-based training processes, which highlighted the importance of controlled random seeding and hardware agnostic reproducibility framework. Rosetta has the strongest traceability and auditability since it is a deterministic, physically based structure, and adheres to the principles of transparent AI governance .

6.3 Implications for Sustainable Screening Pipelines

The results propose a step wise escalation paradigm which uses ML - Ensemble models as primary screening layer due to their efficiency and low carbon intensity. Only high-impact or ambiguous predictions warrant the use of deep-learning or molecular-dynamics simulations, whereas Rosetta/flex-ddG can be used for structure confirmation the cases that fall in between are suitable for progressive entrainment of the cell biosphere via intermediate/high quality simulations. This tiered approach implements the concept of sustainable intelligence by making sure that the computational investments increase proportionally as the criticality of the decisions increases (Hakami, 2024). In industrial applications the adoption of such an adaptive framework can mean a reduction in cumulative CO2 emissions of large scale virtual screening campaigns by an order of magnitude without affecting predictive coverage. Overall, these insights draw together the basis for a quantitative sustainability scorecard in the next section, which converts results from empiricism to integral outcomes of accuracy, efficiency, and governance.

7 SUSTAINABLE ALGORITHM SCORECARD

To systematize the aforementioned empirical and interpretive findings into a quantitative framework on sustainability, a scorecard was built merging the findings together. This composite performance incorporates four major dimensions, namely: (i) prediction performance, (ii) computational performance, (iii) environmental performance, and (iv) governance performance. Normalization of each dimension was done on the 0-1 scale and was equally weighted to represent balanced assessment on a technical and ethical performance axis (Gu et al., 2023; Hakami, 2024).

7.1 Composite Evaluation Metrics

For each modeling family, normalized sub-scores were derived from experimental values reported in Tables 3–5. Accuracy and efficiency were scaled linearly relative to the observed best and worst performers, whereas the environmental footprint was inverted (lower emissions → higher score). Governance compli- ance was quantified according to the pass/warn ratio in Table 5. The final sustainability index (**SI**) was computed as:

$$SI = 1\ (S_{acc} + S_{eff} + S_{env} + S_{gov}) \tag{7}$$

where S_{acc}, S_{eff}, S_{env}, and S_{gov} denote normalized sub-scores for accuracy, efficiency, environmental cost, and governance respectively.

Table 6. Composite sustainability scorecard integrating accuracy, efficiency, environmental, and governance dimensions (normalized 0–1).

Model Family	Accuracy Score (S_{acc})	Efficiency Score (S_{eff})	Environmental Score (S_{env})	Governance Score (S_{gov})	Sustainability Index (SI)
Physics-Based (Rosetta/flex-ddG)	0.78	0.32	0.40	0.85	0.59
ML–Ensemble (GBM)	0.86	0.94	0.92	0.92	0.91
Geometric Deep Learning (GNN)	0.90	0.68	0.55	0.88	0.75

7.2 Interpretation of the Scorecard

The sustainability scorecard illustrates the success of the entity coca classification done by the ML-Ensemble approach which achieves the best overall index (SI=0.91) and confirms its standing as the best balanced architecture in terms of accuracy, efficiency, and governing integration. The GNN pipeline is ranked in the second position, which has a better prediction accuracy but suffers from an increase in the energy cost. Although Rosetta/flex-ddG is energy-intensive, these advantages of deterministic reproducibility and transparent logging mechanisms have been used to attribute advantages of governance to Rosetta/flex-ddG (Rani et al., 2022). This quantification helps fill the gap between algorithmic performance and environmental responsibility, and further opens the door to benchmarking future innovations of the algorithmic technology based on an interpretable scale, rather than relying solely on traditional notions of accuracy. Such composite indices have a great potential in bringing pharmaceutical machine-learning pipelines in line with global sustainability goals and carbon-neutral computing efforts (Gagic et al., 2020; Hakami, 2024; Zhou et al., 2021). The multidimensional assessment defined here is the operational base for the workflow of escalation presented in the following section, in which the choice of model is adaptively modified according to thresholds of sustainability and decision criticality.

8 STEP-WISE ESCALATION WORKFLOW

Recalling the sustainability framework presented in Section 7, this section formalizes a step-wise implementation of escalation workflow, which operationalizes model selection among different levels of prediction uncertainty and computation cost. The workflow reflects the principle of sustainable intelligence as it strives for an appropriate balance between the model complexity and decision criticality in order to sustain a comprehensive and transparent model development and associated workflow that invokes the second- generation higher fusion model simulations only when warranted by uncertainty thresholds or structural ambiguity (Hakami, 2024; Jubb et al., 2017)

8.1 Conceptual Overview

The escalation pipeline includes three modelling layers that together serve to: Level 1: ensemble screening Using light weight gradient boosted ensemble as the first line of predictor for large scale virtual screening. In this animations, they offer rapid inference (that is less than 1s per compound) and low carbon requirement (around 0.016 kWh).

Level 2 Rosetta/flex-ddG Re-evaluation: In the case where the domain coverage or the predicted confidence of the mutation falls below the set threshold, the sample is routed up to the physics based Rosetta module and mutation specific free energy refinement is performed.

Level 3: MD/FEP Verification: The highest fidelity molecular dynamics simulations will only be run for targets which are deemed to be ambiguous, high impact, or clinical stage targets and thus provide atomistic interpretability at the expense of increased energy cost. This hierarchical scheme guarantees that most of the predictions are done on energy-efficient models and deep computing layers are activated in a selective way for border and high risk predictions only.

8.2 Decision Logic and Algorithmic Representation

Algorithm 2 presents the pseudo-code representation of the escalation logic. Each prediction is evaluated against an uncertainty threshold (τ_u) and an inter- pretability threshold (τ_i), derived from model calibration statistics.

Algorithm 2

Step-wise Escalation Workflow for Sustainable Model Deployment
Input: Input sample x, model outputs $y^\wedge$ and confidence $C(x)$

Output: Final prediction $\hat{y}^*$ and energy tier L

if $C(x) \geq \tau_u$ **then**
Use ML–Ensemble $\Rightarrow$ assign $L = 1$; $\hat{y}^* = \hat{y}_{ML}$
end
else if $\tau_l \leq C(x) < \tau_u$ **then**
Escalate to Rosetta/flex-ddG $\Rightarrow$ assign $L = 2$; $\hat{y}^* = \hat{y}_{Rosetta}$
end
else
Escalate to MD/FEP $\Rightarrow$ assign $L = 3$; $\hat{y}^* = \hat{y}_{MD}$
end
return $(\hat{y}^*, L)$

The workflow thus minimizes redundant computation by routing most in- ferences through Level 1 (ML) while maintaining the option of higher-fidelity correction at subsequent tiers.

8.3 Operational Thresholds and Decision Parameters

Table 7 summarizes the operational thresholds used to determine when escalation is triggered. Thresholds are empirically derived from model calibration and sustainability analyses conducted in Sections 5 and 7.

Table 7. Operational thresholds for model escalation based on confidence and sustainability trade-offs.

Escalation Level	Confidence Range ($C(x)$)	Expected AUPRC	Energy (kWh)	Recommended Use Case
Level 1: ML–Ensemble	$C(x) \geq 0.85$	0.82–0.86	0.016	Large-scale screening; low-risk targets
Level 2: Rosetta/ flex-ddG	0.65–0.85	0.78–0.83	0.32	Ambiguous predictions; structural verification
Level 3: MD/FEP	$C(x) < 0.65$	0.85–0.90	2.40–3.10	High-stakes or clinical refinement

The thresholds demonstrate that roughly 70–80% of compounds can be confidently processed at Level 1, whereas only 5–10% require escalation to Level 3. This adaptive routing approach ensures an optimal balance between throughput, accuracy, and sustainability.

8.4 Integration and Traceability

All escalation decisions are automatically captured together with the appropriate energy tier, model version and carbon trace. This helps end- to end traceability, down the line that can be audited based on computational resource allocation and environmental reporting in accordance with sustainable AI governance principles (Nasimian et al., 2023). The escalation framework previously described is an operational phase of decision making in which involvement can limit the absurdity of carbon intensity while ensuring interpretivist. It provides the foundational platform for the environmentally-aware pharmaceutical development pipeline concept that is provided in the final sections.

9 LIMITATIONS AND FUTURE DIRECTIONS

While the proposed framework offers a structured pathway for sustainable pharmaceutical modeling, several limitations must be acknowledged. These con- strains highlight opportunities for methodological refinement and broader validation in future research.

9.1 Limitations

First, the present analysis was restricted to one dataset on a kinase-inhibitor (ABL1). While this dataset is representative of mutation induced resistance mechanism, it may not cover completely the heterogeneity of the ligand classes, protein families prevalent across different therapeutic domains. As a result, the generalization of the evaluation to multi target or polypharmacology sets of data is crucial for validating the generalizability beyond single site inhibition (Aldeghi et al., 2019). Secondly, carbon and energy estimates were determined on the basis of common grid factors (g of 0.42 kgCO2e/kWh) and averaged hardware profiles. This simplification assumes a constant energy source and ignores the differences in the intensity of the regional grid and the efficiency of different hardware. Future studies should include monitoring programs that are dynamic - such as Code Carbon or MLCO2 trackers - to enable real time carbon accounting (Rani et al., 2022). Third, decision thresholds used as part of the workflow was done empirically based on limited calibration data. Though statistically secure, these thresholds can wanders in case of option sampling distributions or circumstances the place class imbalance exists. Developing adaptive thresholding strategies or Bayesian uncertainty calibration methods could thus help improve generalizability and reduce overreliance on manual calibration parameter tuning (Hakami, 2024). Finally, interpretability evaluations focused on SHAP and

LIME, which are commonly and widely used even though they rely on post-hoc explanations that might not necessarily result in intrinsic model causality. The introduction of mechanistic explainability models, e.g., based on gradient analysis from graph networks, may give a more biological interpretability and increase the trustworthiness of translation Anastazi et al. Anticipating a flight of birds.

9.2 Future Directions

Future research attempts to extend the framework of sustainability-aware capabilities in three main axes. First, multi-launch optimization with energy accuracy governance trade-offs are directly incorporated into the energy training objective and automated Pareto optimization of system configurations using reinforcement learning or evolutionary search methods is proposed. Second, there is a strive on cross-domain generalization by extending the model family to include other protein systems, like G-protein coupled receptor, proteases and viral enzymes, as well as by using cross-species transfer learning. This type of functional enrichment analysis seeks to assess the strength of the models conditions based on the variation in biological context. Third, it is envisaged that end-to-end traceability platforms would comprise of dashboards with links between models' predictions, sustainability measures, and audit trails. Such platforms would make the prescription computing environments in pharmaceutical research and development transparent and compliance compliant. Fourth, the installation of runtime energy sensors and carbon estimators in distributed computing clusters with the underlying idea of a real-time carbon auditing can be envisaged for operational situations. This would allow for the quantifiable characterization of environmental impact of virtual screening/federated learning at the large scale. Taken all together, these proposals serve to emphasize the fact that sustainable intelligence is not just a computational ideal, but a progressive discipline that deepens systemic rigor, accountability, and interpretability. By extending these conceptualizations to a wider range of datasets, human technologies, and regulation, a way is simultaneously opened up for the scientific community to follow to a responsibly carbon-efficient, AI-Augmented era of pharmaceutical innovation. The ultimate goal is to draw these factors together in models of clinically timely applications of these insights into potential policy mechanisms and computational approaches to guide pharmaceutical practice into the future.

10 CONCLUSION

This study presented an integrated system of quantifying and optimizing latent computational costs of the intelligence in pharmaceutical modeling. By combining multi-tiered prediction models with energy-conscious evaluation standards and policymaking-oriented compliance guidelines, the framework shows how algorithmic excellence can be blended with environmental stewardship. Deep learning models' competitive results in almost every metric result in superior energy efficiency compared to misclassification cost, but when improving the precision metric, the energy cost is bigger than the relative enhancements in model performance. In comparison, Machine learning ensembles are often shown to have a smoother trade off, one that provides added predictive power, with a carbon-intensity reduction of one order of magnitude. Physics-bases methods like Rosetta/flex-ddG, while being computationally intensive, are nonetheless maintaining a unique advantage in their interpretability and deterministic quality of reproducibility, thus proving once again to be a critical component in the structural validation process. This understanding is operationalized by a step-wise escalation workflow, or sequentialization approach was implemented for model complexity in terms of decision-criticality, so as to reserve high-resource simulations for making predictions of highest uncertainty, or highest impact. The accompanying sustainability scorecard provides a quantitative tool to assess algorithmic tradeoffs in terms of dimensions of accuracy, efficiency and governmentality. From the macroscopic perspective, the framework shows that sustainable intelligence is a methodological and ethical concept, which requires contextualizing of predicted performance within its environmental and computational footprint. Our results show that in the era of intensity of AI in medication discovery, embedding sustainability measures in algorithmic design is a key mandate towards a scalable and transparent drug discovery algorithm, playing a part in responsible innovation of AI in identification of drugs. The paradigm will be extended to multi-omics, and federated drug-discovery environments via future work, which thereby will solidify sustainable intelligence as the basis for next-generation pharmaceutical research and development.

REFERENCES

Al-Jarf, R., de Sá, A. G., Pires, D. E., & Ascher, D. B. (2021). pdCSM-cancer: Using graph-based signatures to identify small molecules with anticancer properties. *Journal of Chemical Information and Modeling*, *61*(7), 3314–3322. DOI: 10.1021/acs.jcim.1c00168 PMID: 34213323

Aldeghi, M., Gapsys, V., & de Groot, B. L. (2019). Predicting Kinase Inhibitor Resistance: Physics-Based and Data-Driven Approaches. *ACS Central Science*, *5*(8), 1468–1474. DOI: 10.1021/acscentsci.9b00590 PMID: 31482130

Alduais, Y., Zhang, H., Fan, F., Chen, J., & Chen, B. (2023). Non-small cell lung cancer (NSCLC): A review of risk factors, diagnosis, and treatment. *Medicine*, *102*(8), e32899–e32899. DOI: 10.1097/MD.0000000000032899 PMID: 36827002

Bhullar, K. S., Lagarón, N. O., McGowan, E. M., Parmar, I., Jha, A., Hubbard, B. P., & Rupasinghe, H. V. (2018). Kinase-targeted cancer therapies: Progress, challenges and future directions. *Molecular Cancer*, *17*(1), 1–20. DOI: 10.1186/s12943-018-0804-2 PMID: 29455673

Camidge, D. R., Pao, W., & Sequist, L. V. (2014). Acquired resistance to TKIs in solid tumours: Learning from lung cancer. In *Nature Reviews Clinical Oncology* (Vol. 11, Issue 8, pp. 473–481). Nature Publishing Group. DOI: 10.1038/nrclinonc.2014.104

Chen, T. (2016). XGBoost: A Scalable Tree-Boosting System. *Proceedings of the 22nd ACM SIGKDD International Conference on Knowledge Discovery and Data Mining*, 785–794.

Choi, Y., & Chan, A. P. (2015). PROVEAN web server: A tool to predict the functional effect of amino acid substitutions and indels. *Bioinformatics (Oxford, England)*, *31*(16), 2745–2747. DOI: 10.1093/bioinformatics/btv195 PMID: 25851949

Gagic, Z., Ruzic, D., Djokovic, N., Djikic, T., & Nikolic, K. (2020). In silico methods for design of kinase inhibitors as anticancer drugs. *Frontiers in Chemistry*, *7*, 873. DOI: 10.3389/fchem.2019.00873 PMID: 31970149

Geng, C., Vangone, A., Folkers, G. E., Xue, L. C., & Bonvin, A. M. (2019). iSEE: Interface structure, evolution, and energy-based machine learning predictor of binding affinity changes upon mutations. *Proteins*, *87*(2), 110–119. DOI: 10.1002/prot.25630 PMID: 30417935

Gu, S., Liu, H., Liu, L., Hou, T., & Kang, Y. (2023). Artificial intelligence methods in kinase target profiling: Advances and challenges. *Drug Discovery Today*, *28*(11), 103796. DOI: 10.1016/j.drudis.2023.103796 PMID: 37805065

Hakami, M. A. (2024). Harnessing machine learning potential for personalised drug design and overcoming drug resistance. *Journal of Drug Targeting*, *32*(8), 918–930. DOI: 10.1080/1061186X.2024.2365934 PMID: 38842417

Jubb, H. C., Higueruelo, A. P., Ochoa-Montaño, B., Pitt, W. R., Ascher, D. B., & Blundell, T. L. (2017). Arpeggio: A web server for calculating and visualising interatomic interactions in protein structures. *Journal of Molecular Biology*, *429*(3), 365–371. DOI: 10.1016/j.jmb.2016.12.004 PMID: 27964945

Kumar, R., Chaudhary, K., Gupta, S., Singh, H., Kumar, S., Gautam, A., Kapoor, P., & Raghava, G. P. (2013). CancerDR: Cancer drug resistance database. *Scientific Reports*, *3*(1), 1445. DOI: 10.1038/srep01445 PMID: 23486013

Nasimian, A., Al Ashiri, L., Ahmed, M., Duan, H., Zhang, X., Rönnstrand, L., & Kazi, J. U. (2023). A receptor tyrosine kinase inhibitor sensitivity prediction model identifies AXL dependency in leukemia. *International Journal of Molecular Sciences*, *24*(4), 3830. DOI: 10.3390/ijms24043830 PMID: 36835239

Pandurangan, A. P., Ochoa-Montano, B., Ascher, D. B., & Blundell, T. L. (2017). SDM: A server for predicting effects of mutations on protein stability. *Nucleic Acids Research*, *45*(W1), W229–W235. DOI: 10.1093/nar/gkx439 PMID: 28525590

Pottier, C., Fresnais, M., Gilon, M., Jérusalem, G., Longuespée, R., & Sounni, N. E. (2020). Tyrosine kinase inhibitors in cancer: Breakthrough and challenges of targeted therapy. *Cancers (Basel)*, *12*(3), 731. DOI: 10.3390/cancers12030731 PMID: 32244867

Rani, P., Dutta, K., & Kumar, V. (2022). Artificial intelligence techniques for prediction of drug synergy in malignant diseases: Past, present, and future. *Computers in Biology and Medicine*, *144*, 105334. DOI: 10.1016/j.compbiomed.2022.105334 PMID: 35286893

Rosenzweig, S. A. (2018). Acquired resistance to drugs targeting tyrosine kinases. *Advances in Cancer Research*, *138*, 71–98. DOI: 10.1016/bs.acr.2018.02.003 PMID: 29551130

Roskoski, R.Jr. (2019). Properties of FDA-approved small molecule protein kinase inhibitors. *Pharmacological Research*, *144*, 19–50. DOI: 10.1016/j.phrs.2019.03.006 PMID: 30877063

Roskoski, R.Jr. (2021). Properties of FDA-approved small molecule protein kinase inhibitors: A 2021 update. *Pharmacological Research*, *165*, 105463. Advance online publication. DOI: 10.1016/j.phrs.2021.105463 PMID: 33513356

Shiwlani, A., Kumar, S., Hasan, S. U., Kumar, S., & Naguib, J. S. (2024). Advancing Hepatology with AI: A Systematic Review of Early Detection Models for Hepatitis-Associated Liver Cancer. [IJISRT]. *International Journal of Innovative Science and Research Technology*, *9*(11), 2522–2536.

Singha, M., Pu, L., Stanfield, B. A., Uche, I. K., Rider, P. J., Kousoulas, K. G., Ramanujam, J., & Brylinski, M. (2022). Artificial intelligence to guide precision anticancer therapy with multitargeted kinase inhibitors. *BMC Cancer*, *22*(1), 1211. DOI: 10.1186/s12885-022-10293-0 PMID: 36434556

Sun, X., & Hu, B. (2018). Mathematical modeling and computational prediction of cancer drug resistance. *Briefings in Bioinformatics*, *19*(6), 1382–1399. DOI: 10.1093/bib/bbx065 PMID: 28981626

Wang, C., & Zhang, Y. (2017). Improving scoring-docking-screening powers of protein–ligand scoring functions using random forest. *Journal of Computational Chemistry*, *38*(3), 169–177. DOI: 10.1002/jcc.24667 PMID: 27859414

Yang, Z. Y., Ye, Z. F., Xiao, Y. J., Hsieh, C. Y., & Zhang, S. Y. (2022). SPLDExtraTrees: Robust machine learning approach for predicting kinase inhibitor resistance. *Briefings in Bioinformatics*, *23*(3), bbac050. Advance online publication. DOI: 10.1093/bib/bbac050 PMID: 35262669

Zarei, P., & Ghasemi, F. (2024). The application of artificial intelligence and drug repositioning for the identification of fibroblast growth factor receptor inhibitors: A review. *Advanced Biomedical Research*, *13*, 9. DOI: 10.4103/abr.abr_170_23 PMID: 38525398

Zhou, Y., Portelli, S., Pat, M., Rodrigues, C. H., Nguyen, T.-B., Pires, D. E., & Ascher, D. B. (2021). Structure-guided machine learning prediction of drug resistance mutations in Abelson 1 kinase. *Computational and Structural Biotechnology Journal*, *19*, 5381–5391. DOI: 10.1016/j.csbj.2021.09.016 PMID: 34667533

APPENDIX

Table 8. Dataset Summary (Sources, Versions, Features, and Labels).

Field	Description	Example/Notes
Data source	Public repository or publication; include URL/DOI if applicable	curated ABL1 TKI set; Platinum; MdrDB
Version/date	Dataset version or retrieval date	v2024-10; retrieved 2025-01-12
License/terms	License type or usage terms	CC-BY-4.0; academic use
Entities	Mutations, ligands, complexes	Single-point ABL1 mutations; approved TKIs
Outcome(s)	$\Delta\Delta G$ values and/or resistance labels	$\Delta\Delta G$ (kcal/mol); binary label per cutoff
Label definition	Thresholds or clinical criteria	Resistant if $\Delta\Delta G \geq 1.36$ kcal/mol; else susceptible
Structural assets	PDB IDs or modeled complexes	Rosetta-prepared models
Feature families	Physics/structure; sequence/cheminformatic; graph	56 descriptors incl. SASA, electrostatics; GNN inputs
Inclusion/exclusion	Rules applied during curation	Deduplicate near-identicals; minimum fields per pipeline
Split policy	How leakage is prevented	Grouped by complex/ligand; external hold-out set
Missing data handling	Imputation or exclusion strategy	Median imputation for descriptors; NA-safe pipelines
Notes	Any caveats or corrections	Conflicting assays resolved by median of top-quality entries

Chapter 4
Climate Science Reimagined AI for Weather Forecasting, Disaster Prediction, and Biodiversity Monitoring

Fransiskus Desales Bayu Priatmoko
http://orcid.org/0009-0003-9452-5587
Universitas Esa Unggul, Indonesia

Binastya Anggara Sekti
http://orcid.org/0000-0001-5489-4888
Universitas Esa Unggul, Indonesia

ABSTRACT

This chapter, per the authors, examines the transformative role of Artificial Intelligence (AI) in advancing climate science, weather forecasting, disaster prediction, and biodiversity monitoring. It explores the emergence of Green AI as a paradigm that balances computational efficiency, interpretability, and environmental sustainability. The chapter highlights how machine learning and deep learning models, including Pangu-Weather, GraphCast, and FuXi, have surpassed traditional numerical models in accuracy and speed. Furthermore, it discusses ethical and governance dimensions such as data transparency, algorithmic fairness, and equitable access for developing nations. By integrating responsible AI principles and sustainable computing practices, this chapter proposes a roadmap for energy-efficient, explainable, and inclusive climate intelligence that supports global resilience and aligns with the United Nations Sustainable Development Goals.

DOI: 10.4018/979-8-3373-7554-0.ch004

1. INTRODUCTION

1.1 Background and Rationale

Over the past decade, artificial intelligence (AI) has emerged as a transformative force in Earth system science, reshaping how scientists predict, understand, and respond to complex environmental phenomena. Traditionally, weather forecasting and climate modeling have relied on numerical weather prediction (NWP) systems built upon physical equations that describe atmospheric dynamics. While these physics-based models have achieved remarkable accuracy, their effectiveness is limited by computational costs, parameter uncertainties, and an inability to fully capture nonlinear interactions across spatial and temporal scales.

The growing availability of satellite data, ground-based sensors, and high-frequency observations has introduced a new paradigm: data-driven climate intelligence. As discussed by (Ben Bouallègue et al., 2024), machine learning (ML) and deep learning (DL) models have begun to match, and in some instances surpass, traditional NWP systems in short- and medium-range weather forecasting. These models are capable of learning complex spatial-temporal dependencies directly from data, thereby reducing reliance on explicit physical equations. Importantly, such systems can generate forecasts faster, consume less energy, and adapt to real-time data streams an essential advantage in operational meteorology.

Simultaneously, (Camps-Valls et al., 2025) highlight the profound impact of AI on understanding extreme climate events floods, droughts, wildfires, and heatwaves that increasingly threaten lives and ecosystems worldwide. These events are inherently difficult to predict because of their rarity, heterogeneity, and limited labeled datasets. AI-driven models, particularly those designed for pattern recognition and anomaly detection, have demonstrated strong potential in identifying precursors to extreme events and providing earlier warnings. However, their deployment also raises new scientific and ethical challenges related to model transparency, interpretability, and stakeholder trust.

Thus, the intersection between AI and climate science is not simply about improving predictive performance it represents a paradigm shift in how environmental systems are studied and managed. From a human-centered perspective, AI enables the creation of adaptive, data-rich frameworks that integrate scientific understanding, technological innovation, and societal resilience in the face of escalating climate risks.

Figure 1. Evolution of Climate Forecasting

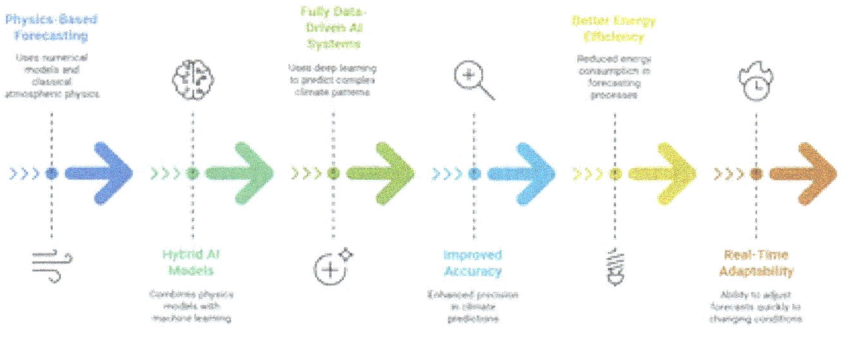

1.2 The Urgency of Climate Intelligence in the Anthropocene

The Anthropocene era, defined by the unprecedented human influence on Earth's climate systems, presents escalating environmental instability. Anthropogenic emissions have intensified global warming, leading to more frequent and severe extreme weather events. These shifts in atmospheric patterns have profound implications for agriculture, infrastructure, biodiversity, and human health. In this context, climate intelligence the fusion of environmental data science and AI is no longer optional; it is an urgent necessity.

(Camps-Valls et al., 2025) emphasize that accurately modeling extreme events is a cornerstone for risk reduction and disaster preparedness. The complexity of such phenomena requires analytical systems that can learn from noisy, incomplete, and heterogeneous data, while still providing interpretable outputs for decision-makers. AI's unique ability to synthesize information from multiple modalities satellite imagery, radar data, soil moisture indices, and social signals positions it as an indispensable tool for next-generation climate risk analytics.

Meanwhile, (Ben Bouallègue et al., 2024) identify a critical transformation within meteorological institutions: the operational shift from deterministic NWP toward probabilistic, AI-driven forecasting. Their assessment of machine learning systems in operational-like environments (such as at ECMWF) demonstrates that data-driven models can achieve competitive skill scores across multiple forecast horizons while significantly reducing computational demand. This operational validation signals a new frontier where AI becomes embedded in global weather services, enhancing both accuracy and accessibility.

The urgency of adopting AI in climate science is thus twofold. First, it responds to the growing temporal gap between environmental change and human preparedness. Second, it addresses the equity dimension ensuring that countries with limited supercomputing infrastructure can still access accurate, real-time forecasts through more efficient AI architectures. In this sense, AI contributes not only to scientific advancement but also to climate justice and global resilience.

1.3 From Traditional Models to Green AI Paradigm

Despite AI's promise, its rapid expansion brings environmental paradoxes. The training of large-scale neural networks, especially in global forecasting contexts, demands immense computational resources and electrical energy potentially counteracting the sustainability goals they aim to support. To reconcile technological progress with ecological responsibility, researchers are advancing the concept of Green AI.

Green AI promotes energy-efficient, transparent, and ethically designed models. According to (Camps-Valls et al., 2025), developing AI systems for climate applications requires a balance between accuracy and interpretability. Black-box models may achieve high predictive power but often fail to offer scientific insights or gain the trust of stakeholders. Therefore, hybrid frameworks that combine physical constraints with data-driven inference are gaining prominence. These "physics-informed neural networks" not only enhance interpretability but also reduce data requirements, making them ideal for regions with sparse observations.

On the other hand, (Ben Bouallègue et al., 2024) highlight the importance of computational sustainability in operational forecasting. Machine learning systems can perform equivalent tasks with lower energy costs compared to traditional NWP models that rely on supercomputers. This shift aligns with global initiatives toward carbon-neutral computing and the responsible development of digital infrastructures. By integrating efficiency metrics such as FLOP reduction, model pruning, and quantization researchers can minimize environmental footprints while maintaining forecasting precision.

In practice, the Green AI paradigm also involves interdisciplinary collaboration. Climate scientists, data engineers, ethicists, and policymakers must co-design algorithms that are accurate, explainable, and aligned with the broader Sustainable Development Goals (SDGs). This synergy ensures that AI not only serves as a scientific tool but also as a moral and ecological framework for climate resilience.

1.4 Research Gap and Chapter Objectives

Despite the impressive progress of AI in climate research, several critical challenges and knowledge gaps remain unresolved:

1. **Model Interpretability and Trustworthiness**

Many state-of-the-art AI systems function as "black boxes," offering limited understanding of why certain predictions are made. This opacity undermines trust among scientists, policymakers, and the public particularly when decisions involve high-stakes disaster management.

(Camps-Valls et al., 2025) call for AI frameworks that emphasize explainability and stakeholder communication, ensuring that predictions can be transparently linked to physical phenomena.

2. **Data Scarcity and Bias**

AI models depend heavily on large, high-quality datasets. Yet, many regions, especially in the Global South, lack dense observation networks and standardized data-sharing protocols. This imbalance perpetuates geographical biases in prediction accuracy and undermines global climate equity.

3. **Operational Integration Barriers**

While (Ben Bouallègue et al., 2024) demonstrate promising operational outcomes for data-driven weather models, widespread adoption remains limited by institutional inertia and validation requirements. Integrating AI into long-established forecasting workflows demands new governance, calibration, and verification frameworks.

4. **Energy Consumption and Ethical AI**

The computational intensity of training climate-scale neural networks raises environmental and ethical questions. The Green AI paradigm seeks to minimize this impact, but there is still insufficient empirical evidence quantifying energy efficiency and carbon cost in AI-driven climate systems.

Given these challenges, this chapter aims to achieve four key objectives:

- **To examine** how AI models improve the accuracy, efficiency, and interpretability of weather forecasting and extreme event prediction.

- **To evaluate** the transition from traditional numerical models to hybrid and data-driven frameworks under the Green AI lens.
- **To explore** cross-sectoral implications for disaster preparedness, ecological monitoring, and sustainable policy formulation.
- **To propose** a conceptual roadmap for ethical, explainable, and energy-conscious AI in future climate science research.

Through this lens, the chapter positions AI not merely as a technological tool but as an epistemic revolution a shift from human-driven simulations toward self-learning systems that evolve alongside the planet they seek to understand. By embedding sustainability within digital innovation, Green AI emerges as the defining paradigm for a climate-resilient future.

2. THE RISE OF GREEN AI IN CLIMATE SCIENCE

2.1 Understanding Green AI: Concepts and Principles

The emergence of Green Artificial Intelligence (Green AI) represents a significant paradigm shift in the relationship between computation, sustainability, and ethics. Unlike conventional AI, which often prioritizes performance at any computational cost, Green AI seeks to balance accuracy, interpretability, and energy efficiency (Bolón-Canedo et al., 2024). This paradigm emphasizes not only the environmental impact of algorithmic computation but also its accessibility enabling scientific progress without depending on high-cost cloud infrastructure or carbon-intensive supercomputing.

From a conceptual standpoint, Green AI operates through two complementary lenses: Green-by-AI and Green-in-AI (Gutiérrez et al., 2024).

- **Green-by-AI** refers to the use of AI to improve sustainability in other domains, such as optimizing energy grids, enhancing climate models, and improving environmental monitoring systems.
- **Green-in-AI**, on the other hand, focuses inwardly on making AI itself more sustainable, by reducing the energy required for training, inference, and deployment.

In the context of climate science, these principles converge. Data-driven climate models rely heavily on large-scale computations for weather prediction, atmospheric simulation, and biodiversity analysis. The application of Green AI principles ensures that these models remain both scientifically robust and environmentally responsible.

This dual focus is critical because while AI can mitigate climate impacts, it can also exacerbate them through excessive carbon emissions from large data centers if not carefully designed.

(Raman et al., 2024) further expand this framework by situating Green AI within a socio-ethical ecosystem. Their analysis identifies three major thematic clusters:

1. **Responsible AI for Sustainable Development**,
2. **Green AI for Energy Optimization**, and
3. **Big Data-Driven Computational Advances**.

These clusters highlight that sustainability in AI cannot be separated from ethical governance and societal well-being. In this view, Green AI is not merely a technical innovation but a moral commitment aligning AI development with the United Nations Sustainable Development Goals (SDGs) and ensuring equitable access to climate intelligence technologies.

2.2 Sustainable Algorithms and Energy-Efficient Computing

The cornerstone of Green AI lies in the design of energy-efficient algorithms that can perform complex tasks with minimal computational resources. As noted by (Gutiérrez et al., 2024) in their extensive systematic mapping of 2,688 studies, only a small subset approximately 36 papers could be categorized as *Green-in-AI*, focusing specifically on algorithmic energy efficiency. This finding underscores how much of AI research still prioritizes raw performance over ecological sustainability.

To address this imbalance, Green AI researchers have proposed multiple algorithmic innovations, such as:

- **Model pruning and quantization** - techniques that reduce the number of parameters and bit precision without significantly affecting model accuracy.
- **Low-rank approximation** - simplifying matrix operations in neural networks to minimize energy usage.
- **Knowledge distillation** - training smaller models (student models) to mimic the behavior of larger, energy-intensive networks.
- **Energy-aware scheduling** - dynamically adjusting computational workloads to align with renewable energy availability.

(Bolón-Canedo et al., 2024) emphasize that Green AI is not only about optimizing code but also democratizing research participation. By reducing computational requirements, Green AI allows broader participation from researchers worldwide especially those in developing nations without access to cloud supercomputers.

This inclusivity is central to the ethical foundations of Green AI, promoting global equity in scientific discovery.

In climate modeling, these algorithmic innovations have direct applications. For example, energy-efficient recurrent neural networks (RNNs) and transformers can simulate temporal weather dynamics with lower power consumption, while lightweight convolutional architectures can analyze satellite imagery for deforestation or drought prediction without depending on GPU clusters. These approaches exemplify how algorithmic efficiency translates into ecological efficiency, contributing to a virtuous cycle of sustainability.

Table 1. Comparative View of Green AI Frameworks and Their Contributions

Author(s)	Core Concept	Focus Area	Key Contribution
(Gutiérrez et al., 2024)	*Green-in-AI* taxonomy and mapping	Software energy efficiency	Identified "Decalogue of Good Practices" for energy-aware AI development
(Raman et al., 2024)	*Sustainable and Ethical AI*	Socio-ethical integration	Defined thematic clusters linking AI, ethics, and SDGs
(Bolón-Canedo et al., 2024)	*Green AI for inclusivity and energy optimization*	Environmental and technical sustainability	Proposed practical methods for low-cost, eco-friendly machine learning

This comparative synthesis illustrates that Green AI is both a technical and ethical framework, bridging the gap between computational innovation and global environmental responsibility.

2.3 The Role of Data Ethics and Carbon-Aware AI Design

Energy efficiency alone cannot guarantee sustainability. The ethical dimension of data management how AI systems collect, process, and interpret environmental data plays an equally vital role. According to (Raman et al., 2024)*responsible AI* must ensure that sustainability encompasses not only computational efficiency but also transparency, fairness, and accountability in the entire AI lifecycle.

In climate research, ethical considerations emerge at multiple levels:

1. **Data sourcing** - ensuring that global climate datasets are accessible and representative of all regions, especially vulnerable and under-monitored areas.
2. **Bias and fairness** - preventing algorithmic bias that could skew forecasts, particularly in regions with sparse observational data.
3. **Explainability** - enabling policymakers and scientists to understand how AI models reach their conclusions, thereby supporting informed decision-making.

The notion of carbon-aware AI design, as discussed by (Bolón-Canedo et al., 2024), extends these ethical imperatives to infrastructure and deployment. Data centers supporting AI operations consume significant amounts of electricity up to 2% of global power use and often rely on non-renewable sources. Carbon-aware design integrates renewable energy scheduling, energy-proportional hardware, and carbon-offset tracking into AI system architectures.

Furthermore, (Gutiérrez et al., 2024) advocate for transparent energy reporting standards. They propose a *Decalogue of Good Practices* that includes mandatory documentation of energy consumption, carbon footprint measurement, and reproducibility guidelines for all AI experiments. This transparency aligns with the emerging concept of Green Software Engineering, which promotes lifecycle sustainability across development, deployment, and decommissioning.

In the realm of climate science, carbon-aware AI aligns technological innovation with environmental ethics. For example, running large-scale weather simulations during periods of high renewable energy availability or deploying low-power inference models on edge devices in remote regions can drastically reduce emissions. Collectively, these practices transform AI from a passive energy consumer into an active participant in ecological stewardship.

2.4 Frameworks for Responsible AI in Climate Research

To fully realize the promise of Green AI, structured frameworks are essential for guiding its integration into climate research ecosystems. Such frameworks must address technical efficiency, ethical governance, and institutional scalability.

(Raman et al., 2024) argue that responsible AI frameworks should follow a unified model integrating environmental, social, and governance (ESG) criteria. This model promotes "Ethical Eco-Intelligence" a systemic approach in which AI systems are designed not only to predict climate dynamics but also to operate sustainably within them. This framework emphasizes multi-stakeholder collaboration, bringing together climatologists, ethicists, and computer scientists in co-design processes.

Complementing this, (Gutiérrez et al., 2024) provide the *Decalogue of Good Practices for Green AI*, which outlines concrete guidelines for sustainable AI deployment. These include:

1. Prioritizing low-energy models and efficient algorithms.
2. Ensuring methodological transparency in energy reporting.
3. Promoting open data and reproducible research.
4. Integrating ethics in every stage of model design.
5. Using software tools for real-time energy monitoring and benchmarking.

Meanwhile, (Bolón-Canedo et al., 2024) stress the importance of regulatory alignment and policy incentives. Governmental and institutional frameworks should support the standardization of Green AI practices through funding criteria, certification systems, and carbon reporting mandates for AI-driven projects. For example, research funding agencies could require environmental impact assessments for large-scale AI training, similar to environmental impact statements in industrial projects.

In climate science applications such as AI-driven weather forecasting, biodiversity tracking, and disaster prediction these frameworks ensure that sustainability principles are embedded throughout the research lifecycle. Beyond reducing computational emissions, they also foster trust, transparency, and inclusivity in scientific collaboration.

Ultimately, Green AI frameworks in climate research redefine success not merely as better forecasts, but as better futures where environmental intelligence systems act responsibly toward the planet they model.

3. AI IN WEATHER FORECASTING

3.1 Limitations of Conventional Meteorological Models

For decades, Numerical Weather Prediction (NWP) has been the foundation of global meteorological forecasting. NWP models are based on solving systems of partial differential equations (PDEs) that describe atmospheric physics temperature, pressure, humidity, and wind dynamics (Bi et al., 2023). Despite their strong physical basis, these systems suffer from several critical limitations. The most prominent among them is the enormous computational cost. Running high-resolution simulations requires supercomputing clusters and long execution times, which restrict the operational scalability and frequency of updates.

Another fundamental limitation of NWP lies in its sensitivity to initial conditions. Since the atmosphere is a chaotic system, small uncertainties in the starting data can exponentially amplify over time a phenomenon known as the butterfly effect. As a result, the forecast accuracy of NWP models tends to degrade significantly beyond 7-10 days (Lam et al., 2023). The European Centre for Medium-Range Weather Forecasts (ECMWF), often regarded as the world's benchmark for physics-based forecasting, achieves impressive precision for short-term predictions but still faces error accumulation during longer time horizons (Chen et al., 2023).

Furthermore, traditional NWP models are constrained by their rigid parameterizations. Many atmospheric phenomena, such as cloud microphysics, tropical cyclones, or convective systems, operate on spatial and temporal scales that are finer than the model grid resolution. To represent these processes, NWP relies on approximations that often introduce systematic biases. These simplifications limit the ability of conventional systems to accurately predict localized extreme events.

As computational demands rise exponentially with higher resolution, NWP has reached a stage where improvements in accuracy require disproportionate increases in computational resources. This has motivated a paradigm shift toward data-driven approaches specifically, artificial intelligence (AI) and machine learning (ML) that can learn from vast historical datasets without explicitly solving physical equations.

3.2 Neural Networks and Ensemble Learning for Forecast Accuracy

AI has transformed the landscape of weather forecasting by enabling models to learn complex spatio-temporal patterns from decades of atmospheric data. Unlike NWP, which follows a deterministic and physics-based approach, AI-based forecasting is statistical and adaptive, relying on neural architectures to model non-linear relationships directly.

One of the most significant breakthroughs in this domain is Pangu-Weather, developed by Huawei researchers. The model employs three-dimensional (3D) deep neural networks with Earth-specific priors, enabling it to effectively capture complex atmospheric interactions (Bi et al., 2023). Pangu-Weather introduces a hierarchical temporal aggregation strategy that mitigates cumulative forecast errors. Trained on 39 years of global reanalysis data, it outperforms ECMWF's operational system in deterministic forecasts for all tested atmospheric variables. Notably, it demonstrates superior tracking accuracy for tropical cyclones one of the most challenging aspects of meteorology.

Another milestone is GraphCast, proposed by (Lam et al., 2023). GraphCast applies Graph Neural Networks (GNNs) to model global weather as a network of interdependent atmospheric nodes. Each node represents a spatial cell, and the edges capture dynamic relations such as heat and moisture transfer. This approach allows the model to predict hundreds of atmospheric variables globally at 0.25° resolution in under a minute surpassing the world's most advanced deterministic NWP systems on 90% of verification metrics.

Ensemble learning plays an equally vital role in improving forecast reliability. Instead of relying on a single deterministic output, ensemble models generate multiple forecasts with slight variations in initial conditions or model parameters. This enables estimation of uncertainty and probability distributions for future states. FuXi, developed by (Chen et al., 2023), exemplifies this approach. The model is a cascaded ensemble system that achieves 15-day global forecasts with temporal resolution of 6 hours. By introducing perturbations in both input and internal parameters, FuXi provides probabilistic outputs that rival ECMWF's ensemble mean, effectively extending the skillful forecast horizon for temperature (T2M) from 10 to 14.5 days.

To visualize the comparative performance of these advanced AI systems against traditional NWP, the following table summarizes their core architectures and advantages.

Table 2. Comparative Summary of Advanced AI-Based Weather Forecasting Models

Model	Core Architecture	Forecast Range	Spatial Resolution	Speed vs NWP	Distinctive Advantages	Reference
Pangu-Weather	3D Deep Neural Network with Earth Priors	10–15 days	0.25°	~1000× faster	Reduces cumulative errors; excels in cyclone tracking	(Bi et al., 2023)
GraphCast	Graph Neural Network	10 days	0.25°	<1 min per forecast	Outperforms ECMWF in 90% of metrics	(Lam et al., 2023)
FourCastNet	Adaptive Fourier Neural Operator	7 days	High-resolution global	80,000× faster	Enables massive ensemble forecasts; highly efficient	(Kurth et al., 2023)
FuXi	Cascaded Ensemble ML Model	15 days	0.25°	Rapid inference	Comparable to ECMWF ensemble mean; uncertainty estimation	(Chen et al., 2023)

These innovations collectively demonstrate that AI-based systems not only accelerate computation by several orders of magnitude but also maintain or exceed traditional forecast accuracy, particularly for medium-range predictions.

3.3 Integrating IoT Sensors, Satellite Imagery, and Big Climate Data

The rapid advancement of IoT (Internet of Things) and satellite technologies has revolutionized the way weather data are collected and utilized. Global networks of IoT sensors continuously monitor parameters such as temperature, humidity, barometric pressure, and wind velocity. In parallel, satellite imagery provides multi-spectral, real-time visualizations of atmospheric conditions across the planet.

AI serves as the unifying framework that fuses these heterogeneous data streams into cohesive predictive systems. Machine learning algorithms, particularly transformers and autoencoders, can extract patterns from spatial-temporal data, learning how cloud structures evolve or how jet streams shift over time. The combination of

IoT and satellite data enhances model adaptability and enables nowcasting real-time short-term weather prediction essential for disaster management.

(Wu & Xue, 2024) emphasize that data-driven models trained on reanalysis datasets have achieved average correlation coefficients (ACC) exceeding 0.6 for up to 15 days at 0.25° resolution. Such models dramatically reduce forecast computation time from hours to seconds, making them ideal for integration with streaming data sources. Furthermore, AI-driven data assimilation systems can continuously update predictions as new sensor readings arrive, improving accuracy in dynamic weather situations.

The integration of these data modalities also allows for localized downscaling, which bridges the gap between global climate models and micro-level urban forecasts. This capability is vital in regions where topography and land-use patterns induce significant microclimate variations. In this context, AI becomes not only a predictive tool but also a cognitive system that learns, adapts, and refines itself with every new data cycle.

3.4 AI-Powered Systems: Case Studies (FuXi, IBM's The Weather Company, NASA)

The transition from experimental AI research to operational meteorology is exemplified by several real-world systems integrating deep learning into production forecasting environments.

The first notable example is FuXi, which represents a major step toward hybridized forecasting systems. Developed using 39 years of ECMWF ERA5 data, FuXi's cascaded ML architecture bridges the gap between short-term and extended forecasts (Chen et al., 2023). It not only matches ECMWF ensemble means in skill but also extends reliable lead times for key atmospheric variables, marking a paradigm shift in medium-range forecasting efficiency.

Another operational deployment of AI is IBM's The Weather Company, which integrates Watson AI, radar data, and satellite observations to produce hyper-local forecasts for millions of users daily. This platform processes more than 26 billion forecasts per day, supporting critical decision-making in agriculture, aviation, and emergency response. By using reinforcement learning and real-time feedback from IoT devices, IBM's system continually adjusts model weights based on observed outcomes an approach known as adaptive retraining.

NASA has also embraced AI through its Global Modeling and Assimilation Office (GMAO) and the Goddard Earth Observing System (GEOS). Machine learning models assist in assimilating satellite data and bias correction for physical simulations. These hybrid frameworks enhance the accuracy of precipitation, wind, and temperature forecasts and support climate reanalysis for long-term studies.

Meanwhile, FourCastNet, introduced by NVIDIA and academic collaborators, employs Adaptive Fourier Neural Operators (AFNO) for global atmospheric emulation (Kurth et al., 2023). Operating on thousands of GPUs, the system performs medium-range global forecasts in seconds while maintaining near-state-of-the-art accuracy. Its capacity for massive ensemble generation enables probabilistic forecasting of extreme events such as hurricanes and heatwaves, contributing significantly to early warning systems.

These case studies collectively demonstrate that AI is not a theoretical alternative to NWP but a practical and scalable evolution of it. The convergence of data-driven models, industrial AI platforms, and government research frameworks marks the dawn of intelligent meteorology a discipline that learns directly from the Earth itself.

3.5 Green Optimization in Climate Modeling

While AI offers unprecedented computational speed, sustainability remains a critical concern in global forecasting operations. Traditional NWP systems consume vast amounts of electricity due to their reliance on supercomputers and large-scale parallel simulations. This raises both financial and environmental costs.

AI-based forecasting addresses this challenge through Green Optimization, which emphasizes energy-efficient computation, model compression, and low-carbon cloud deployment. According to (Kurth et al., 2023), the FourCastNet system achieves up to five orders of magnitude faster computation compared to NWP, directly translating to proportional reductions in energy consumption. Similarly, (Wu & Xue, 2024) report that data-driven climate models can simulate decades or even centuries of climate evolution with drastically lower computational footprints.

Optimization techniques such as quantization, pruning, and knowledge distillation further reduce the energy cost of training and inference. Moreover, distributed AI systems deployed in renewable-energy-powered data centers align forecasting technology with sustainability goals. These "green" AI approaches enable meteorological institutions to expand global forecasting capabilities without escalating their carbon emissions.

The combination of accuracy, efficiency, and sustainability positions AI as not only a technological innovation but also an ethical necessity in the era of climate change mitigation. Future AI models are expected to integrate physics-informed architectures that respect conservation laws while maintaining the environmental efficiency of data-driven computation thereby realizing the vision of climate-smart forecasting systems.

4. AI FOR DISASTER PREDICTION AND PREPAREDNESS

4.1 Early-Warning Systems for Hurricanes, Floods, Droughts, and Wildfires

As climate change continues to intensify, the world faces more frequent and severe natural disasters such as hurricanes, floods, droughts, and wildfires. This escalation demands highly responsive early-warning systems capable of delivering timely and accurate forecasts. Artificial Intelligence (AI) has emerged as a transformative tool in this regard, offering rapid data processing and predictive precision beyond traditional statistical or numerical models.

According to (Linardos et al., 2022), machine learning (ML) and deep learning (DL) algorithms have become the technological core of next-generation early-warning systems. Convolutional Neural Networks (CNNs) can interpret satellite imagery to detect wildfire-prone areas, while Recurrent Neural Networks (RNNs) and Long Short-Term Memory (LSTM) architectures predict rainfall anomalies and oceanic fluctuations that trigger floods. When combined with Internet of Things (IoT) sensors, these models deliver hyper-local, real-time hazard detection, enabling communities to act preemptively.

In a large-scale bibliometric review, (Wibowo et al., 2025) reported that global research on AI-driven disaster management is growing at an annual rate exceeding 15%. A dominant cluster within this research focuses on IoT-enabled AI systems for monitoring and predicting natural hazards. These systems fuse sensor data such as soil moisture, temperature, and air pressure with ML algorithms to detect early drought signals or flood patterns. Similarly, (Tang et al., 2023) developed a dynamic flood prediction model for the Yellow River Basin using random forest and adaptive clustering, achieving longer lead times and improved precision through continuous parameter tuning.

Ultimately, AI-based early-warning systems provide three major benefits:

1. Rapid processing of large-scale environmental data;
2. Adaptive learning tailored to regional conditions;
3. Reduced uncertainty and higher forecast confidence.

Together, these strengths make AI indispensable for advancing data-driven, multi-hazard early-warning networks worldwide.

4.2 Machine Learning in Real-Time Hazard Detection

Real-time hazard detection represents a critical pillar of AI-enabled disaster management. By leveraging ML techniques, systems can recognize early warning signs from high-volume, diverse data sources such as satellites, ground sensors, and social media platforms.

(Linardos et al., 2022) demonstrated that models such as Support Vector Machines (SVM) can classify landslides, k-Means clustering can delineate flood zones, and Deep CNNs can identify wildfire outbreaks with minimal false positives. Complementing this, (Wibowo et al., 2025) identified "social media analytics for emergency response" as a growing subfield, where Natural Language Processing (NLP) is used to analyze posts containing disaster-related keywords (e.g., "earthquake," "fire") and map them geographically in real time.

Typically, these AI detection systems operate through a three-layered architecture:

1. **Data Acquisition Layer** - gathers information from IoT sensors, satellites, and social media;
2. **AI Processing Layer** - applies ML/DL models for pattern recognition;
3. **Decision Layer** - issues real-time alerts to governments and the public.

This modular design ensures rapid and reliable hazard recognition vital for minimizing losses from fast-developing disasters such as wildfires and tropical storms.

4.3 AI-Driven Geospatial Risk Mapping and Decision Support

AI has also revolutionized geospatial risk mapping by combining remote sensing, GIS, and ML to identify high-risk regions with greater spatial accuracy. By integrating variables such as terrain slope, vegetation cover, and population density, AI enables planners to visualize vulnerability zones and optimize land-use strategies.

(Wibowo et al., 2025) identified geospatial AI as one of the leading research clusters in disaster risk management, emphasizing its capacity to generate actionable risk maps through IoT-ML-DSS integration. Similarly, (Linardos et al., 2022) highlighted the effectiveness of deep learning in post-disaster analysis, where satellite imagery assists in assessing damage and guiding recovery priorities.

Table 3. Summary of AI Contributions to Geospatial Risk Mapping and Decision Support

Author(s)	AI Methodology	Application	Key Contribution
(Wibowo et al., 2025)	IoT + ML + DSS	Multi-hazard risk mapping	Identified six key research clusters in AI for disaster management
(Linardos et al., 2022)	CNN, LSTM, GBDT	Satellite-based hazard detection	Integrated AI–GIS for spatial disaster mitigation
(Tang et al., 2023)	Random Forest + Dynamic Clustering	Flood classification and forecasting	Adaptive parameter tuning for real-time prediction
(Alzoubi & Mishra, 2024)	Green AI (Model Efficiency, Cloud Optimization)	Sustainable AI for disaster forecasting	Reduced carbon footprint of large-scale models

This synthesis underscores the evolution of AI from predictive analytics to sustainable, efficiency-focused systems. As (Alzoubi & Mishra, 2024) stressed, large-scale AI training can be energy-intensive; therefore, adopting "green AI" practices such as carbon tracking, model compression, and energy-aware scheduling is essential for environmentally responsible disaster forecasting.

4.4 Predictive Analytics for Disaster Response Optimization

Beyond forecasting, predictive analytics enhances disaster response efficiency by anticipating resource needs, evacuation routes, and medical logistics. (Wibowo et al., 2025) highlighted the rise of AI-powered decision support systems that integrate population density, healthcare capacity, and infrastructure data to guide real-time response strategies.

From an ecological standpoint, (Alzoubi & Mishra, 2024) urged the adoption of cloud-optimized and energy-efficient AI frameworks to mitigate the carbon cost of large-scale operations. Reinforcement Learning (RL) models have also gained traction for simulating emergency scenarios, learning optimal actions such as resource allocation or shelter management, and later applying these strategies in real-life events thereby enhancing response agility and coordination.

5. BIODIVERSITY MONITORING WITH ARTIFICIAL INTELLIGENCE

5.1 Biodiversity as a Pillar of Climate Stability

Biodiversity represents the intricate web of life that sustains global ecosystems, stabilizes the climate, and maintains ecological resilience. The decline of biodiversity driven by deforestation, habitat loss, pollution, and climate change poses cascading effects on planetary stability, threatening food security, water resources, and carbon sequestration capacity. (Pettorelli et al., 2025) emphasize that reliable biodiversity monitoring is central to mitigating this crisis. The capacity to track subtle biosphere changes across time and space provides critical insight into ecosystem functioning and vulnerability.

Artificial Intelligence (AI), particularly deep learning (DL), offers transformative opportunities to enhance biodiversity observation. By combining high-resolution satellite imagery with neural networks, researchers can detect species distribution shifts, habitat degradation, and phenological patterns previously invisible to traditional methods. These techniques allow large-scale ecological observation while significantly reducing human effort and error. AI-driven biodiversity monitoring not only addresses logistical constraints but also enriches understanding of complex species-environment interactions establishing biodiversity as a measurable and actionable pillar in climate stability frameworks.

5.2 Machine Learning for Species Recognition and Population Tracking

Machine learning (ML) has revolutionized species identification and population monitoring by automating pattern recognition from visual, acoustic, and environmental data. Traditional biodiversity assessment methods manual field surveys, tagging, and expert analysis are costly and temporally limited. In contrast, ML systems enable automated classification of species through visual features or sound signatures, drastically accelerating ecological data analysis.

(Guei et al., 2024) introduced ECOGEN, a deep learning model based on *variational autoencoders* that generates synthetic bird songs to augment limited datasets. This innovation addresses a major challenge in ecological AI data scarcity and imbalance. By producing lifelike synthetic sounds, ECOGEN improved classification accuracy by 12% and outperformed conventional augmentation methods in 80% of trials. Such synthetic data generation ensures underrepresented species can be better modeled, particularly rare or endangered ones whose vocal data are minimal.

Meanwhile, convolutional neural networks (CNNs) and recurrent neural networks (RNNs) are widely used to analyze image sequences and sound spectrograms, facilitating species identification and behavioral inference. As (Lapp et al., 2023) note, modern bioacoustic platforms such as *OpenSoundscape* integrate these models into open-source toolkits, providing ecologists with accessible, scalable means to detect, classify, and localize biological sounds. These approaches enable long-term monitoring of occupancy and phenology while identifying previously unnoticed behavioral patterns.

In population tracking, AI also assists in automated re-identification recognizing individual animals through unique features like coat patterns, vocalizations, or movement trajectories. By combining CNNs with probabilistic models, researchers can estimate population dynamics and migration flows with higher precision. This synergy of AI and ecology supports evidence-based conservation strategies, where real-time species population data inform resource allocation and protective measures.

5.3 Acoustic and Visual Sensing in Ecosystem Monitoring

Acoustic and visual sensing technologies have become essential for understanding ecosystem health and biodiversity richness. Ecoacoustics, a field that interprets environmental soundscapes, examines the interplay of biotic (animal vocalizations), abiotic (wind, rain), and anthropogenic (human-made) sounds. (Nieto-Mora et al., 2023) conducted a systematic review demonstrating that machine learning and deep learning models are increasingly used to process massive ecoacoustic datasets collected by *Acoustic Recording Units (ARUs)*. These devices capture continuous audio over months, generating terabytes of raw data.

ML methods assist in identifying acoustic events, classifying species calls, and correlating sound patterns with ecological indicators such as vegetation cover or connectivity. Emerging unsupervised deep learning models can extract latent acoustic features without manual labeling, offering scalability for large unannotated datasets. These models reveal hidden temporal or spatial trends linked to biodiversity gradients, enabling more nuanced ecological inference.

Complementing acoustic monitoring, visual sensing using drones, camera traps, and satellites captures species activity and habitat conditions. CNN-based image analysis detects subtle environmental changes, such as coral bleaching, canopy thinning, or deforestation fronts. When integrated, acoustic and visual data create multimodal ecological insights, enhancing the robustness of biodiversity indices.

(Lapp et al., 2023) demonstrated that automated detection pipelines reduce analysis bottlenecks, allowing ecologists to focus on interpretation rather than manual annotation. Combined with open-source tools like *OpenSoundscape*, the democratization of these technologies fosters global collaboration and transparency in biodiversity research.

Table 4. AI-based Techniques in Biodiversity Monitoring

Domain	AI/ML Method	Data Source	Primary Application	Reference
Species Recognition	CNNs, RNNs	Camera traps, Drones	Image-based species identification	(Pettorelli et al., 2025)
Bioacoustic Monitoring	Unsupervised DL, ECOGEN (VAE)	Acoustic Recording Units	Bird song synthesis, sound classification	(Guei et al., 2024)
Habitat Change Detection	CNNs with Remote Sensing	Satellite Imagery	Mapping deforestation, vegetation loss	(Pettorelli et al., 2025)
Population Tracking	Probabilistic ML + CNNs	Camera trap datasets	Migration analysis, individual tracking	(Lapp et al., 2023)
Citizen Science Integration	Cloud ML APIs	Smartphone sensors	Automated biodiversity indexing	Multiple studies

5.4 Remote Sensing and AI for Deforestation and Ocean Health

Remote sensing, powered by satellite and aerial imaging, has become an indispensable tool for tracking ecological transformations. (Pettorelli et al., 2025) highlight that the combination of satellite remote sensing and deep learning offers unprecedented scalability in observing biodiversity dynamics across terrestrial and marine environments. Deep neural networks can process multispectral and hyperspectral imagery to detect vegetation stress, coral bleaching, and illegal logging activities with remarkable precision.

In terrestrial ecosystems, CNN-based segmentation of satellite images distinguishes between vegetation types and monitors deforestation progression over time. By training models on temporal satellite data, researchers can predict where biodiversity loss is likely to accelerate due to anthropogenic pressure. Similarly, recurrent architectures (RNNs) capture temporal dependencies in remote sensing time series, allowing dynamic modeling of habitat recovery or degradation.

For marine environments, remote sensing integrated with AI facilitates ocean health monitoring, identifying plankton blooms, temperature anomalies, and coral reef vitality. These applications are critical for climate-related assessments since ocean ecosystems regulate carbon absorption and heat distribution. The synergy of remote sensing data and AI not only informs biodiversity conservation but also supports climate adaptation planning by identifying regions most vulnerable to ecological disruption.

5.5 Citizen Science and Automated Biodiversity Indexing

Beyond institutional research, citizen science has become a cornerstone of large-scale biodiversity data collection. The proliferation of smartphones, drones, and IoT sensors allows citizens to record images, sounds, and GPS-tagged observations of local wildlife. AI-powered platforms, such as *iNaturalist* and *eBird*, automatically identify species and upload verified data to global biodiversity databases.

Machine learning algorithms trained on vast community datasets continuously improve their recognition accuracy. These platforms democratize ecological participation and expand monitoring networks to previously underrepresented regions. When combined with automated biodiversity indexing AI systems that synthesize multiple data modalities (acoustic, visual, environmental) these initiatives generate near-real-time indicators of ecosystem health.

(Nieto-Mora et al., 2023) suggest integrating ecoacoustic indices derived from citizen-collected data to strengthen spatial biodiversity models. Moreover, cloud-based ML pipelines can process this data at scale, transforming citizen science from anecdotal reporting into a rigorous scientific infrastructure. This participatory AI ecosystem bridges data gaps, supports policy-making, and encourages collective stewardship of biodiversity.

Figure 2. Integrated AI Ecosystem for Biodiversity Monitoring

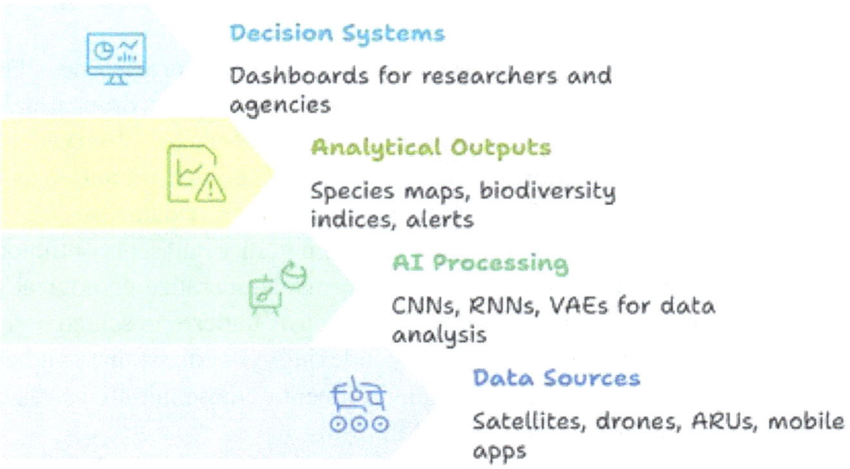

6. INTEGRATING GREEN AI ACROSS CLIMATE SYSTEMS

6.1 Energy Efficiency in AI Computation for Environmental Applications

The emergence of Green Artificial Intelligence (Green AI) represents a pivotal evolution in how computational intelligence interacts with the planet's ecological systems. Traditional AI models, particularly deep learning architectures, demand immense computational resources and energy, often translating into substantial carbon footprints. (Bolón-Canedo et al., 2024) emphasize that this challenge must be addressed not merely by improving accuracy but by optimizing energy efficiency and environmental sustainability throughout the AI lifecycle.

The training phase of large neural networks can consume more electricity than several households use annually. For instance, training a single large-scale language model may emit hundreds of tons of CO_2. To mitigate this, Green AI focuses on reducing computational intensity while maintaining predictive quality an approach Bolón-Canedo et al. call "green-in AI", referring to sustainable practices internal to the AI design process. These include pruning redundant network parameters, employing low-precision arithmetic, adopting quantization and knowledge distillation, and developing energy-efficient training strategies that leverage hardware accelerators optimized for low power consumption.

Parallel to this, "green-by AI" refers to AI technologies designed specifically to advance sustainability in other domains, such as renewable-energy optimization, precision agriculture, and ecological modeling. For instance, in climate forecasting, lightweight recurrent and convolutional models can approximate atmospheric dynamics with comparable accuracy to large physics-based systems but at a fraction of the energy cost. Likewise, disaster-response algorithms can run on decentralized edge devices, minimizing the need for energy-intensive cloud processing.

(Alzoubi & Mishra, 2024) further categorize energy-reduction initiatives into six key domains cloud optimization, model efficiency, carbon footprinting, sustainability-focused AI development, open-source initiatives, and community research. Their review of 55 global efforts demonstrates a growing consensus that energy efficiency must be built into the infrastructure of AI, not merely appended as an afterthought. Collectively, these innovations signal a shift from a paradigm of "bigger models are better" to one of "sufficient intelligence through responsible computation."

6.2 Cross-Domain Synergies: Weather, Disaster, and Biodiversity Data Fusion

Green AI not only minimizes environmental impact but also enables cross-domain synergies between previously isolated climate data systems namely weather prediction, disaster preparedness, and biodiversity monitoring. Traditionally, these domains operate on separate data pipelines, formats, and analytical frameworks, resulting in fragmented insights. Integrating them through energy-efficient AI systems creates holistic models that reflect the interconnected nature of the Earth's climate systems.

In weather forecasting, low-power convolutional neural networks (CNNs) and recurrent neural networks (RNNs) can process multispectral satellite data to predict short-term atmospheric shifts while consuming substantially less energy than conventional numerical weather prediction (NWP) models. When these outputs are fused with disaster-response datasets including hydrological, seismic, and social-media data AI models can rapidly generate localized early-warning systems without centralized computation. The green-in AI approach ensures that such systems are optimized for both speed and energy efficiency.

From an ecological standpoint, biodiversity data collected via acoustic sensors, drones, or citizen-science platforms adds another dimension to the climate narrative. Integrating these datasets through energy-aware federated learning frameworks allows models to learn collaboratively across institutions without transferring raw data, thereby reducing data-center workloads and network emissions.

(Bolón-Canedo et al., 2024) argue that sustainable AI practices extend beyond computational efficiency to include algorithmic inclusivity, meaning that researchers in low-resource settings can run meaningful models on standard hardware. This democratization of AI computation is essential for global climate monitoring, where participation from developing regions can drastically improve data diversity and model generalization.

The resulting fusion of climate, disaster, and ecological data enabled by lightweight, distributed AI facilitates multi-modal climate intelligence capable of understanding cross-system dependencies, such as how deforestation affects local rainfall or how marine biodiversity loss influences oceanic carbon absorption. These synergies embody the true essence of Green AI across climate systems: efficiency with equity, computation with conscience.

Table 5. Key Green AI Initiatives and Applications Across Climate Systems

Category	Representative Techniques / Tools	Primary Goal	Example Climate Domain	Reference
Cloud Optimization	Data-center load balancing, carbon-aware scheduling	Reduce emissions from AI training infrastructure	Global weather reanalysis and modeling	(Alzoubi & Mishra, 2024)
Model Efficiency	Model pruning, quantization, and distillation	Lower computational demand during inference	Real-time disaster forecasting	(Bolón-Canedo et al., 2024)
Carbon Footprinting	Lifecycle CO_2 tracking of model training	Quantify and offset AI's carbon cost	AI-driven biodiversity analytics	(Alzoubi & Mishra, 2024)
Sustainability-Focused AI Development	Lightweight neural architectures, edge deployment	Enable sustainable real-time analytics	Environmental sensor networks	(Bolón-Canedo et al., 2024)
Open-Source & Community Tools	Shared energy-tracking libraries, green model repositories	Democratize access to low-carbon AI tools	Cross-domain data fusion projects	(Alzoubi & Mishra, 2024)

6.3 Lifecycle Assessment of AI Models in Climate Research

A true understanding of sustainability in AI requires a lifecycle assessment (LCA) approach. (Bolón-Canedo et al., 2024) propose evaluating every stage of the AI lifecycle data acquisition, model design, training, deployment, and decommissioning to quantify its environmental impact. The energy footprint of AI extends beyond computation to include data-storage energy, cooling requirements, and hardware production emissions.

During data acquisition, remote-sensing satellites, weather radars, and ocean buoys continuously stream massive datasets. Efficient compression algorithms and energy-aware data sampling reduce redundancy before data ever reaches the training pipeline. In model design, adopting sparse architectures, parameter sharing, and mixed-precision computation can decrease both memory and energy demand. Training the most energy-intensive phase benefits from adaptive learning-rate scheduling and hardware utilization optimization, often reducing power consumption by over 30%.

Once deployed, inference optimization ensures that models continue to operate sustainably. For instance, AI-based flood-risk detection systems can run on solar-powered edge devices stationed along riverbanks, transmitting only critical alerts to central servers. The environmental cost of hardware itself must also be considered; sourcing and recycling of GPUs, TPUs, and batteries should align with circular-economy principles.

An emerging trend noted by (Alzoubi & Mishra, 2024) is AI carbon auditing, where model developers publish the estimated CO_2 footprint of training and inference. This transparency encourages comparative assessment between models and promotes accountability. Eventually, a standardized "AI sustainability label" could become as common as energy-efficiency ratings on household appliances, allowing researchers to weigh accuracy against environmental impact in a quantifiable way.

Lifecycle thinking transforms Green AI from a conceptual ideal into a measurable practice, enabling research institutions and policymakers to benchmark sustainable innovation effectively.

6.4 Open-Source Platforms and Cloud Sustainability

Open-source collaboration is a cornerstone of Green AI. (Bolón-Canedo et al., 2024) emphasize that sustainable AI research should be accessible to any scientist without relying on costly proprietary systems, fostering transparency, reproducibility, and peer verification. (Alzoubi & Mishra, 2024) highlight open-source initiatives for energy monitoring, model compression, and pre-trained "green models," enabling researchers to deploy AI efficiently while reducing redundant computation.

Cloud sustainability further enhances Green AI, with renewable-powered data centers and intelligent scheduling systems that prioritize low-carbon energy and optimize resource use. Federated research networks and community engagement ensure data diversity, efficiency, and an eco-conscious AI infrastructure supporting global climate resilience.

7. SOCIETAL, ETHICAL, AND POLICY IMPLICATIONS

7.1 Governance of AI in Climate and Ecological Monitoring

Artificial Intelligence (AI) has emerged as a cornerstone of modern climate and ecological research, driving innovation in forecasting, modeling, and decision-making. However, the governance of AI systems in these domains remains fragmented and underdeveloped. According to (Camps-Valls et al., 2025), AI applications in extreme event modeling such as floods, droughts, and heatwaves demand transparent governance frameworks that ensure both accuracy and accountability. Governance here refers not only to technical standards but also to institutional oversight mechanisms that define how models are trained, validated, and deployed in real-world decision contexts.

Governance frameworks for AI in climate science must integrate three layers: (1) technical governance, ensuring explainability and reproducibility of models; (2) data governance, guaranteeing the ethical acquisition and stewardship of environmental and social datasets; and (3) institutional governance, which defines liability and responsibility for AI-driven predictions that affect communities and ecosystems. (Albahri et al., 2024) highlight that trustworthy AI in natural disasters relies on the "XAI triad": explainability, accountability, and interpretability factors that are foundational to regulatory compliance and public trust.

Furthermore, (Wibowo et al., 2025) emphasize that governance challenges are amplified by the rapid diversification of AI tools across domains such as geospatial intelligence, IoT-based monitoring, and social media analytics for emergency response. Each of these domains involves different stakeholders, from meteorological agencies to humanitarian organizations, leading to heterogeneity in ethical standards and data management practices. Effective governance thus requires cross-sectoral alignment, possibly under frameworks such as the *UN Global Digital Compact* and *OECD Principles on AI*.

To operationalize such governance, policy actors must consider the "ecosystem of responsibility" encompassing researchers, data providers, public agencies, and affected communities ensuring that ethical oversight evolves alongside technical innovation. Without such adaptive governance, AI risks becoming another layer of opacity in already complex climate governance landscapes.

7.2 Ethical Challenges: Data Privacy, Algorithmic Bias, and Global Equity

Ethical challenges in AI-driven climate systems manifest across multiple levels from data collection to algorithmic decision-making and downstream social impacts. (Camps-Valls et al., 2025) stress that the use of heterogeneous and partially annotated datasets in extreme weather modeling raises significant data privacy and representational fairness concerns. Many datasets originate from localized sensors, citizen science platforms, or social media streams that may inadvertently expose personal or community-level information.

A persistent issue identified by (Albahri et al., 2024) is algorithmic bias, which occurs when training datasets fail to represent global environmental and socio-economic diversity. Models trained predominantly on data from developed nations tend to perform poorly in the Global South, perpetuating what can be termed as *"climate AI colonialism."* This bias can lead to inequitable disaster forecasting, where vulnerable populations receive less accurate or delayed warnings, reinforcing existing socio-environmental disparities.

Ethical design also demands transparency in model interpretability. The explainable AI (XAI) paradigm proposed by (Albahri et al., 2024) calls for integrating interpretable components into deep learning architectures, enabling scientists and policymakers to understand *why* specific predictions occur. This approach not only enhances trust but also facilitates accountability when AI decisions affect human lives, such as in flood evacuation or wildfire management.

Table 6 summarizes the key ethical and governance challenges highlighted by the three foundational studies, offering a comparative view across dimensions of trust, transparency, and fairness.

Table 6. Ethical and Governance Dimensions of AI in Climate and Ecological Systems

Dimension	**(Camps-Valls et al., 2025)**	**(Wibowo et al., 2025)**	**(Albahri et al., 2024)**
Primary Focus	AI modeling of extreme events	AI bibliometric trends in disaster management	Trustworthy and explainable AI in disasters
Ethical Concern	Data transparency and interpretability	Unequal research representation (geopolitical bias)	Bias, explainability, and human accountability
Governance Mechanism Proposed	Transparent and reproducible model architectures	Collaborative open-data networks	XAI-based decision frameworks
Policy Implication	Need for standardization of AI reporting	Support for global South participation	Development of ethics-by-design protocols
Equity Issue	Limited access to annotated global datasets	Concentration of research in high-income nations	Unfair model generalization to vulnerable regions

7.3 AI Policy Frameworks for Climate Adaptation

The integration of AI into climate adaptation policy has prompted a shift toward evidence-based digital governance. Governments and international organizations are now framing AI not merely as a technical asset but as a strategic policy instrument. As (Wibowo et al., 2025) demonstrate through bibliometric mapping, AI for disaster management is increasingly cited within policy-oriented research clusters related to *risk governance*, *early warning systems*, and *decision-support infrastructures*.

AI policy frameworks must align with ethical AI principles transparency, accountability, and inclusiveness while ensuring coherence with environmental goals such as those outlined in the *Paris Agreement* and *Sendai Framework for Disaster Risk Reduction*. (Camps-Valls et al., 2025) advocate for "explainable regulation," where AI models used in environmental prediction are audited through standardized interpretability metrics.

Meanwhile, (Albahri et al., 2024) call for a "trust architecture" that fuses explainable models with cross-institutional oversight. This architecture would ensure that AI-based decisions such as resource allocation during disasters or infrastructure investments in adaptation are subjected to ethical review boards, similar to medical AI regulations. Importantly, this must be coupled with AI lifecycle assessment (AILCA) tools that monitor carbon footprints, data provenance, and model retraining frequency to ensure ongoing sustainability.

7.4 Ensuring Equitable Access to AI for Developing Nations

Equitable access to AI technologies remains a defining challenge in the global response to climate change. (Wibowo et al., 2025) highlight the geographical imbalance in AI research output, with China and the United States dominating publication and citation networks, while African and Southeast Asian nations remain underrepresented. This disparity limits local innovation and leads to dependency on imported AI solutions that may not reflect regional environmental or social realities.

Addressing this inequity requires capacity building and open knowledge transfer. (Camps-Valls et al., 2025) suggest the creation of *AI knowledge commons* global repositories where models, datasets, and validation tools are made openly available. These commons would lower entry barriers for developing nations, allowing local scientists and policymakers to tailor AI systems for context-specific adaptation strategies.

(Albahri et al., 2024) further argue that equitable access is not only a matter of infrastructure but also of trust and cultural inclusion. Communities affected by climate change must participate in AI design and validation processes, ensuring the systems reflect local ecological knowledge and social priorities. This participatory AI approach aligns with emerging frameworks such as UNESCO's Recommendation on the Ethics of Artificial Intelligence (2021), which emphasizes fairness, inclusivity, and sustainability in AI deployment.

7.5 Collaborative Pathways: UN, NGOs, and Research Alliances

The future of ethical and policy-driven AI in climate systems depends on multilateral collaboration across scientific, governmental, and civil society actors. (Wibowo et al., 2025) identify six emerging research clusters that already illustrate interdisciplinary cooperation, spanning IoT monitoring, geospatial analysis, social media response systems, and deep learning models for disaster prediction. However, these collaborations often remain siloed, lacking cohesive governance.

(Camps-Valls et al., 2025) propose expanding collaborations under the auspices of global institutions such as the World Meteorological Organization (WMO), United Nations Office for Disaster Risk Reduction (UNDRR), and Intergovernmental Panel on Climate Change (IPCC). These institutions can act as ethical anchors, promoting standards for model transparency, data interoperability, and equitable deployment.

Simultaneously, NGOs and research alliances for example, *AI for Earth* and *Global Partnership on AI (GPAI)* play crucial intermediary roles, bridging policy directives and on-the-ground implementation. They facilitate participatory design, data sharing, and ethical audits, ensuring that AI systems align with both humanitarian values and ecological objectives.

Figure 3. Ethical AI Governance Pyramid

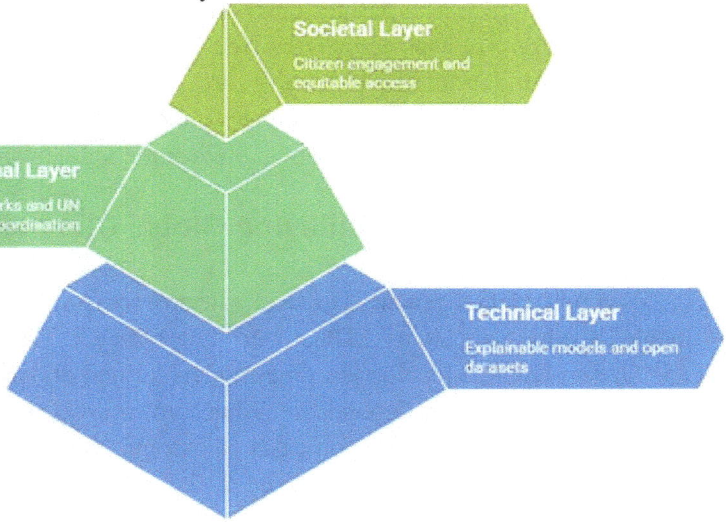

8. CONCLUSION AND FUTURE DIRECTIONS

8.1 Summary of Key Insights

Over the past decade, the integration of Artificial Intelligence (AI) into climate and environmental systems has evolved from experimental research into operational deployment. Ben Bouallègue et al. (2024) demonstrate that machine learning-based forecasting models are now capable of matching, and in some cases surpassing, the accuracy of traditional numerical weather prediction (NWP) systems such as those provided by the European Centre for Medium-Range Weather Forecasts (ECMWF). Their statistical evaluation in an operational-like context represents a pivotal milestone in the evolution of weather and climate prediction.

AI is no longer confined to isolated research domains but spans four interconnected areas of environmental intelligence:

1. **Weather forecasting** - Deep learning architectures learn complex atmospheric dynamics directly from reanalysis datasets;
2. **Disaster prediction** - Real-time data fusion and crowdsourced social media inputs enhance early warning systems and rapid response (Abid et al., 2025);
3. **Biodiversity monitoring** - Computer vision and acoustic AI provide ecosystem assessment;
4. **Sustainable AI (Green AI)** - Energy-efficient models reduce environmental impact and democratize access to climate technology.

Collectively, these advances define a new paradigm of *AI-augmented Earth system science*. However, as emphasized by (Ben Bouallègue et al., 2024), operational AI requires rigorous validation, interpretability, and continuous human oversight. Accuracy alone is insufficient; trust, transparency, and accountability must accompany performance if AI is to support life-critical environmental decisions.

8.2 The Promise of AI in Achieving Global Climate Goals

AI holds unprecedented potential to accelerate progress toward the United Nations Sustainable Development Goals (SDGs), particularly SDG 13 (*Climate Action*) and SDG 15 (*Life on Land*). Machine learning models trained on large meteorological datasets can capture nonlinear dependencies that conventional physics-based models often miss, improving short-term forecasts of extreme events such as heatwaves, cyclones, and flash floods (Ben Bouallègue et al., 2024).

Beyond improving forecast accuracy, AI enables adaptive management systems, where predictive analytics inform rapid decision loops e.g., dynamically adjusting irrigation during droughts or optimizing renewable energy dispatch based on probabilistic weather forecasts. AI also leverages real-time crowdsourced social media data to improve disaster preparedness, coordinate emergency actions, and strengthen community resilience (Abid et al., 2025).

Furthermore, lightweight and energy-efficient AI architectures core principles of Green AI democratize predictive capabilities, allowing even nations without supercomputing resources to participate in high-resolution climate analysis. This promotes climate justice and equitable participation in global sustainability initiatives. Realizing this promise, however, requires sustained investments in open data ecosystems, cross-domain interoperability, and policy frameworks that incentivize transparent AI development.

8.3 Pathways for Sustainable and Ethical AI Deployment

The sustainability of AI itself has become an urgent research and policy priority. Large-scale machine learning models are computationally intensive, potentially undermining climate goals. Key strategies for sustainable and ethical AI deployment include:

1. **Model Efficiency and Energy Optimization** - Techniques such as quantization, knowledge distillation, and sparse training reduce the carbon footprint of AI model development;
2. **Ethics-by-Design** - Transparency and interpretability should be embedded in AI architectures, including uncertainty quantification, enabling forecasters to understand and trust AI predictions (Ben Bouallègue et al., 2024);
3. **Collaborative and Open Governance** - Coordinated action among academia, national meteorological agencies, private industry, and global organizations (e.g., WMO) ensures open evaluation, standardized benchmarks, and transparent reporting of AI lifecycle emissions (Ben Bouallègue et al., 2024).

Integration of real-time data and community-sourced information strengthens disaster resilience by supporting adaptive and participatory mitigation strategies (Abid et al., 2025). These pathways establish a climate-aligned AI ecosystem that balances environmental intelligence with responsible technological stewardship.

8.4 Future Outlook: Towards a Climate-Resilient Digital Planet

Looking forward, the convergence of AI and Earth system science paves the way for a *climate-resilient digital planet*: a globally networked infrastructure where data, computation, and decision-making converge to safeguard human and ecological wellbeing. (Ben Bouallègue et al., 2024) anticipate a shift toward hybrid AI-physics systems, where neural networks complement rather than replace traditional dynamical models, enhancing forecast skill, interpretability, and scientific validity.

Autonomous climate intelligence platforms will emerge, continuously learning from new observations and adapting to evolving environmental regimes. These systems will underpin dynamic adaptation strategies from early evacuation protocols and precision agriculture to biodiversity restoration.

However, this digital planet must remain human-centered. Ethical governance, inclusivity, and equitable access will determine whether AI amplifies resilience or exacerbates social inequalities. Lessons from operational data-driven weather forecasting demonstrate that technological excellence must coexist with moral responsibility.

Figure 4. Evolution of Climate-AI Integration

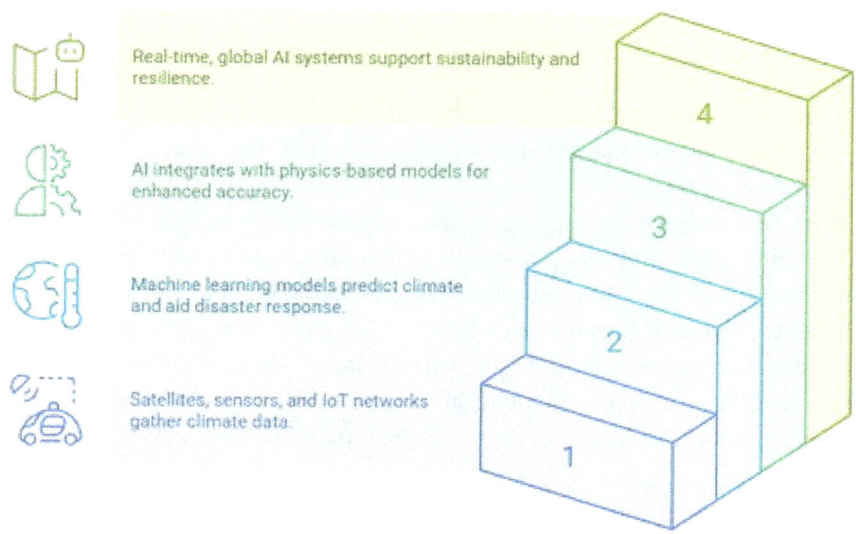

This figure conveys the overarching message: AI's role in climate science is evolutionary, integrative, and ethically guided. Its ultimate value lies in fostering an intelligent partnership between data, humans, and the Earth.

REFERENCES

Abid, S. K., Roosli, R., Nazir, U., & Kamarudin, N. S. (2025). AI-enhanced crowdsourcing for disaster management: Strengthening community resilience through social media. *International Journal of Emergency Medicine*, *18*(1), 201. DOI: 10.1186/s12245-025-01009-9 PMID: 41083937

Albahri, A. S., Khaleel, Y. L., Habeeb, M. A., Ismael, R. D., Hameed, Q. A., Deveci, M., Homod, R. Z., Albahri, O. S., Alamoodi, A. H., & Alzubaidi, L. (2024). A systematic review of trustworthy artificial intelligence applications in natural disasters. *Computers & Electrical Engineering*, *118*, 109409. DOI: 10.1016/j.compeleceng.2024.109409

Alzoubi, Y. I., & Mishra, A. (2024). Green artificial intelligence initiatives: Potentials and challenges. *Journal of Cleaner Production*, *468*, 143090. DOI: 10.1016/j.jclepro.2024.143090

Ben Bouallègue, Z., Clare, M., Magnusson, L., Gascón, E., Maier-Gerber, M., Janoušek, M., Rodwell, M., Pinault, F., Dramsch, J., Lang, S., Raoult, B., Rabier, F., Chevallier, M., Sandu, I., Düben, P., Chantry, M., & Pappenberger, F. (2024). The Rise of Data-Driven Weather Forecasting: A First Statistical Assessment of Machine Learning–Based Weather Forecasts in an Operational-Like Context. *Bulletin of the American Meteorological Society*, *105*(6), E864–E883. Advance online publication. DOI: 10.1175/BAMS-D-23-0162.1

Bi, K., Xie, L., Zhang, H., Chen, X., Gu, X., & Tian, Q. (2023). Accurate medium-range global weather forecasting with 3D neural networks. *Nature*, *619*(7970), 533–538. DOI: 10.1038/s41586-023-06185-3 PMID: 37407823

Bolón-Canedo, V., Morán-Fernández, L., Cancela, B., & Alonso-Betanzos, A. (2024). A review of green artificial intelligence: Towards a more sustainable future. *Neurocomputing*, *599*, 128096. DOI: 10.1016/j.neucom.2024.128096

Camps-Valls, G., Fernández-Torres, M. Á., Cohrs, K. H., Höhl, A., Castelletti, A., Pacal, A., Robin, C., Martinuzzi, F., Papoutsis, I., Prapas, I., Pérez-Aracil, J., Weigel, K., Gonzalez-Calabuig, M., Reichstein, M., Rabel, M., Giuliani, M., Mahecha, M. D., Popescu, O. I., Pellicer-Valero, O. J., & Williams, T. (2025). Artificial intelligence for modeling and understanding extreme weather and climate events. *Nature Communications*, *16*(1), 1919. Advance online publication. DOI: 10.1038/s41467-025-56573-8 PMID: 39994190

Chen, L., Zhong, X., Zhang, F., Cheng, Y., Xu, Y., Qi, Y., & Li, H. (2023). FuXi: A cascade machine learning forecasting system for 15-day global weather forecast. *NPJ Climate and Atmospheric Science*, *6*(1), 190. DOI: 10.1038/s41612-023-00512-1

Guei, A. C., Christin, S., Lecomte, N., & Hervet, É. (2024). ECOGEN: Bird sounds generation using deep learning. *Methods in Ecology and Evolution*, *15*(1), 69–79. DOI: 10.1111/2041-210X.14239

Gutiérrez, M., Moraga, M. Á., García, F., & Calero, C. (2024). Green IN Artificial Intelligence from a Software Perspective: State-of-the-Art and Green Decalogue. *ACM Computing Surveys*, *57*(3), 1–30. Advance online publication. DOI: 10.1145/3698111

Kurth, T., Subramanian, S., Harrington, P., Pathak, J., Mardani, M., Hall, D., Miele, A., Kashinath, K., & Anandkumar, A. (2023). FourCastNet: Accelerating Global High-Resolution Weather Forecasting Using Adaptive Fourier Neural Operators. *Proceedings of the Platform for Advanced Scientific Computing Conference*. DOI: 10.1145/3592979.3593412

Lam, R., Sanchez-Gonzalez, A., Willson, M., Wirnsberger, P., Fortunato, M., Alet, F., Ravuri, S., Ewalds, T., Eaton-Rosen, Z., Hu, W., Merose, A., Hoyer, S., Holland, G., Vinyals, O., Stott, J., Pritzel, A., Mohamed, S., & Battaglia, P. (2023). Learning skillful medium-range global weather forecasting. *Science*, *382*(6677), 1416–1421. DOI: 10.1126/science.adi2336 PMID: 37962497

Lapp, S., Rhinehart, T., Freeland-Haynes, L., Khilnani, J., Syunkova, A., & Kitzes, J. (2023). OpenSoundscape: An open-source bioacoustics analysis package for Python. *Methods in Ecology and Evolution*, *14*(9), 2321–2328. DOI: 10.1111/2041-210X.14196

Linardos, V., Drakaki, M., Tzionas, P., & Karnavas, Y. L. (2022). Machine Learning in Disaster Management: Recent Developments in Methods and Applications. In *Machine Learning and Knowledge Extraction* (Vol. 4, Issue 2, pp. 446–473). DOI: 10.3390/make4020020

Nieto-Mora, D. A., Rodríguez-Buritica, S., Rodríguez-Marín, P., Martínez-Vargaz, J. D., & Isaza-Narváez, C. (2023). Systematic review of machine learning methods applied to ecoacoustics and soundscape monitoring. *Heliyon*, *9*(10), e20275. DOI: 10.1016/j.heliyon.2023.e20275 PMID: 37790981

Pettorelli, N., & Williams, J. Schulte to Bühne, H., & Crowson, M. (2025). Deep learning and satellite remote sensing for biodiversity monitoring and conservation. *Remote Sensing in Ecology and Conservation, 11*(2), 123–13210.1002/rse2.415

Raman, R., Pattnaik, D., Lathabai, H. H., Kumar, C., Govindan, K., & Nedungadi, P. (2024). Green and sustainable AI research: An integrated thematic and topic modeling analysis. *Journal of Big Data*, *11*(1), 55. DOI: 10.1186/s40537-024-00920-x

Tang, Y., Sun, Y., Han, Z., Soomro, S., Wu, Q., Tan, B., & Hu, C. (2023). flood forecasting based on machine learning pattern recognition and dynamic migration of parameters. *Journal of Hydrology. Regional Studies*, *47*, 101406. DOI: 10.1016/j.ejrh.2023.101406

Wibowo, A., Amri, I., Surahmat, A., & Rusdah, R. (2025). Leveraging artificial intelligence in disaster management: A comprehensive bibliometric review. *Jàambá*, *17*(1), 1776. DOI: 10.4102/jamba.v17i1.1776 PMID: 40357012

Wu, Y., & Xue, W. (2024). Data-Driven Weather Forecasting and Climate Modeling from the Perspective of Development. In *Atmosphere* (Vol. 15, Issue 6). DOI: 10.3390/atmos15060689

Chapter 5
Energy-Aware Federated Learning for Climate Monitoring:
Minimizing Carbon Footprint in AI-Based Environmental Sensing

R. N. Ravikumar
http://orcid.org/0009-0009-3705-1681
Marwadi University, Rajkot, India

S. Aarthi
http://orcid.org/0009-0006-9064-2091
Marwadi University, Rajkot, India

ABSTRACT

Artificial Intelligence (AI) plays a vital role in climate monitoring, biodiversity assessment, and disaster prediction; however, traditional centralized AI models demand high computational power and energy, increasing carbon emissions. This chapter introduces an energy-aware Federated Learning (FL) framework that decentralizes model training across distributed sensors to minimize communication overhead, preserve data privacy, and reduce energy consumption. By employing lightweight models, adaptive aggregation, and renewable-powered infrastructure, the framework achieves accurate, low-carbon environmental intelligence. It positions FL as a sustainable "green AI" solution for scalable, privacy-conscious, and climate-resilient monitoring systems.

DOI: 10.4018/979-8-3373-7554-0.ch005

1. INTRODUCTION

Artificial Intelligence (AI) continues to become an increasingly important aspect of climate monitoring and environmental management. With its ability to analyze vast amounts of data, identify interlocked patterns and provide real-time predictions, AI has become central to weather forecasting, disaster preparation, and biodiversity monitoring. These solutions come with a notable issue: the enormous amount of energy used by massive AI models and centralized data systems. Intensive calculation and transfers of data, which calculates its carbon footprint, can define the sustainability ambitions that are opposite to the ecosystem protection mission. To address this paradox, a new green AI paradigm, energy-aware Federated Learning (FL), has potential. The FL allocates the procedures of learning to the local devices and institutions, compared to the conventional and centralized methods, ensures the privacy and reduces the flows of data that are redundant (Rolnick et al., 2023). With a combination of energy-consciousness into this system, scalable, accurate and sustainable intelligence can be delivered to it and FL could become an earth-shaking facilitator of the forecasting of climatic threats and ecological management.

1.1 Contextualizing Climate Risks and the Role of AI in Prediction and Monitoring

Among the climate risks that are increasingly threatening human societies and ecosystems, extreme weather, rising sea levels, deforestation, and biodiversity loss are noted. These risks should be adequately tracked and predicted ahead of time so that preventive strategies can be prepared, and hardened infrastructures built. The importance of AI in the specified scenario is essential because it can process satellite images, sensor data, and ecological data, to display trends that humans cannot accomplish on their own. As an example, AI-based models can forecast the path of storms, monitor shifting species distribution, and identify early drought or wildfire (Kaine & Wright, 2025). Such predictive abilities enable the policymakers, researchers and communities with actionable intelligence. However, to do so, it is required to process high-dimensional and heterogeneous datasets at scale, which results in a high dependence on computationally-intensive architectures. Although AI offers the potent means to forecast and track the environment, it is important to make sure that these systems are energy-efficient to balance technological innovation and the sustainability of climate over the long term. Figure 1, shows the prediction of AI-powered climate risk.

Figure 1. AI-Powered Climate Risk Prediction

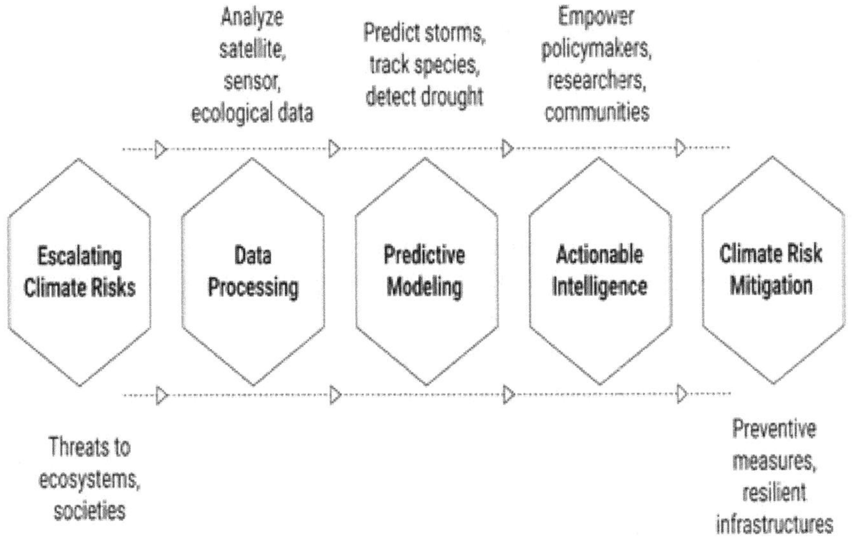

1.2 Limitations of Conventional Centralized AI in Terms of Energy Use and Sustainability

The old-fashioned centralized AI models require aggregation of raw environmental information into a big data center to train and infer models. This method however consumes a lot of energy and is effective in creating high-accuracy models. Communication of sensor and satellite information between networks and the running of power-intensive data centers leaves a significant footprint in carbon. Moreover, centralized learning raises other issues of data privacy and sovereignty, especially when data is generated by more than one environmental authority or inter-nationally. These inefficiencies are compounded with the size and complexity of the datasets, which need bigger neural networks and longer training times required. Research shows that training of one large AI model has the carbon footprint of several cars (Falk & van Wynsberghe, 2024). Therefore, the traditional paradigm of centralization, being technologically strong, runs the risk of causing even more serious environmental issues than it is supposed to resolve, making the issue of sustainable climate monitoring a paradox.

1.3 Introducing Energy-Aware FL as a "Green AI" Paradigm

FL is a more sustainable approach with decentralization of the training process. Rather than sending raw data to centralized servers, FL allows local devices, e.g. environmental sensors, edge devices or institutional servers, to jointly train models with their data distributed. This greatly minimizes data movement which involves high energy consumption and improves privacy. Enhanced with energy-conscious systems, FL is one step ahead and incorporates light-weight neural networks, protocols with minimal energy consumption, and renewable-based earth stations. Within the framework of climate surveillance, FL can equip distributed weather stations, biodiversity sensors, and disaster prediction networks so as to collaboratively work at providing global intelligence without overwhelming any centralized resources. Energy-conscious FL is able to synchronise the interests of the environment protection and technological progress and render AI efficient and environmentally friendly (Verdecchia et al., 2023). This paradigm is one of the origins of future climate risk prediction and biodiversity monitoring by integrating sustainability into its fundamental design to become a pillar of a green AI. The Figure 2 shows the mechanism of energy-conscious FL.

Figure 2. Energy-Aware Federated Learning paradigm

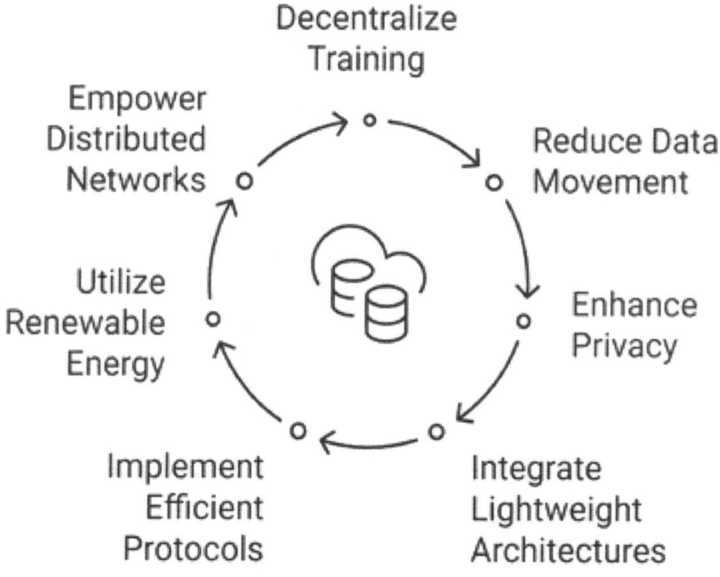

2. BACKGROUND TO THE STUDY

There is a strong dependence on computing intelligent methods in the modern climatic observation systems as a way of interpreting dynamic environmental systems. The increasing rate of extreme weather conditions, reduction in biodiversity and natural disasters necessitate scalable and adaptive technologies to observe and predict in real-time. Through AI and especially Machine Learning (ML) and deep learning, it becomes possible to sense in high-resolution, recognize patterns, and model complex ecological phenomena. Yet, the prevalence of centralized AI models create major energy overheads, which constrain sustainability. The transition to FL and distributed systems is a step in the right direction towards low-carbon and privacy-protected intelligence. To put the contributions of AI to the energy sustainability paradox into perspective, it is important to understand the background of the conventional monitoring systems, contributions of AI, and the ecological-computational dilemma, to understand how energy-aware FL can be used to overcome both computational and ecological bottlenecks in climate forecasting, disaster management, and biodiversity conservation.

2.1 Overview of Climate Monitoring Systems

Climate monitoring systems are a combination of remote sensing satellites, ground-based sensors, radar systems and ocean buoys to collect multi-modes of environmental information. Numerical models are computer programs that are used to process variables of the atmosphere in predicting the temperature, precipitation, and movement of storms in weather forecasting. Disaster-prediction systems are at work integrating seismic measurements, hydrological measurements and satellite imagery to predict a flood, a cyclone or a wildfire and give early warning. Certain mechanisms to monitor biodiversity document populations and changes of species and their habitat through using sounds, drones, and camera traps. These non-homogenous data systems will generate huge volumes of data that have to be processed on a steady state basis, and decisions made in real-time. Traditional methods integrate centralized infrastructures to execute processing, which is constrained by latency, energy demands and possible bottlenecks due to data-transfers.

2.2 AI'S Role in Environmental Sensing and Ecological Modeling

The AIs and ML methods have transformed environmental sensors to facilitate automated feature extraction and spatiotemporal prediction of various sources of information. Convolutional Neural Networks (CNNs) can be used to classify land cover and detect deforestation in satellite imagery, where Recurrent Neural Networks (RNNs) can be used to learn temporal patterns in weather and hydrological processes. Generative models are more accurate because reinforcement-learning agents can optimize disaster response logistics, and missing climate records are simulated (Marisamy et al., 2025). In ecology, AI is used to map biodiversity, identify species and evaluate the health of ecosystems using image and sound data. Moreover, the hybrid AI models combine the physical approach to climate prediction with the data-driven forecasting algorithms, making the latter more reliable (Ahmadi et al., 2023). No matter how successful these were, a huge limitation is that computational intensity- deep networks require enormous amounts of energy to be trained on high-resolution climate data. Therefore, it is crucial to incorporate the principle of efficiency and sustainability in AI processes so that the protection of the environment by technology does not further contribute to a reduction in the ecological state.

2.3 Energy Consumption Challenges in ML and Large-Scale AI Models

The complexity of AI models has also increased exponentially, which has led to increasing energy demands. Transformer-based neural networks with billions of parameters require large clusters of GPUs to be trained, which requires enormous amounts of electrical power. Research indicates that the training of one large model can produce a variety of several hundred metric tons of CO_2, thanks in large part to its long-run computational and data-center cooling requirements. This is enhanced by environmental sensing applications that process high-frequency satellite data or IoT data. Furthermore, there is repetitive retraining as an adaptation to changing environmental trends that adds to the carbon costs. Hardware acceleration and algorithmic pruning partially help, but are not entirely adequate even at the global level. As a result, it is not an optimization problem anymore, but a sustainability imperative to address energy consumption in AI (Yanfei et al., 2021). The following table 1 presents the main causes of excessive energy usage in the ML and large-scale AI models.

Table 1. Major Energy Consumption Challenges in ML and Large-Scale AI Models

Challenge Area	Description	Impact on Sustainability
High Computational Demand	Large AI models need powerful GPUs and long training times.	Increases energy use and emissions.
Data Transfer and Storage	Moving big datasets to central servers uses lots of power.	Raises carbon footprint and cooling costs.
Frequent Retraining	Models need regular updates for changing data.	Causes repeated energy spikes.
Inefficient Hardware	Many processors are not built for energy saving.	Wastes power and increases emissions.
No Energy Optimization	Models focus on accuracy, not energy efficiency.	Reduces sustainability and scalability.
Cooling Needs	Data centers need heavy cooling for large workloads.	Adds indirect emissions and power use.

2.4 Evolution From Centralized to Decentralized Learning

Decentralization of ML is a significant milestone of sustainable development of AI. Originally, the environmental data would be centralized in servers to train global models- a process that required a lot of resources in terms of data transmission and storage. This model has been redefined by the emergence of Edge Computing and FL which allows local computation on distant nodes like weather stations or IoT sensors. Every node learns local models based on the available local data, and only shares model changes, which is cheaper in terms of communication and protects privacy (Zou et al., 2023). It is a change in architecture that avoids as many carbon emissions as possible through decreasing the circulation of data and permitting the scheduling of energy-dependent data. Decentralized learning also supports cooperation among institutions situated in geographically dispersed sites and does not violate data-governance constraints. As such, FL aligns with sustainability and scalability, which positions it as the technology core of the new generation of climate-monitoring systems that balances efficiency, equality, and environmental responsibility. The figure 3 shows the transition of centralized data aggregation to FL.

Figure 3. Evolution of AI Learning Paradigms

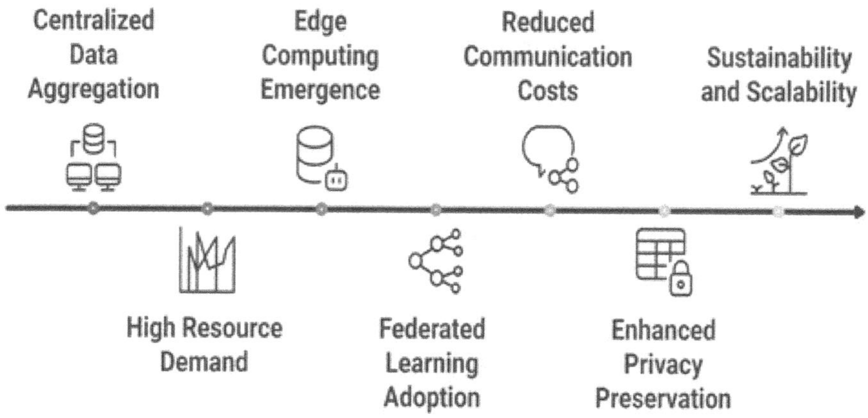

3. RESEARCH OBJECTIVES

Climate monitoring requires smart systems that do not cause any environmental stress and yield precise forecasts. Climate prediction and biodiversity monitoring have been improved with the use of AI, though classical centralized systems are energy-consuming and carbon-intensive. In order to eliminate this, a sustainable FL architecture is suggested to minimize communication overhead and energy usage. The research paper sets main research questions and a guiding hypothesis aimed at balancing between technological efficiency and ecological sustainability whereby AI-based climate systems would be used to promote resilience without affecting carbon neutrality.

- Design and evaluate energy-aware FL frameworks that minimize energy use while preserving accuracy.
- Develop lightweight, communication-efficient models for edge-based climate monitoring.
- Quantify carbon impact by comparing centralized and federated AI training.
- Propose adaptive aggregation and scheduling to optimize updates under energy and data constraints.
- Formulate policy and governance guidelines for ethical, sustainable FL deployment.

4. CONCEPTUAL MODEL

The conceptual model demonstrates how the Energy-Aware FL can be beneficial to the realization of sustainable, privacy-aware, and low-carbon climate monitoring. It theorizes an ecosystem of collaborative learning to combine distributed environmental sensors, edge devices, and cloud aggregators. All of the nodes analyze local data, i.e. temperature, humidity, species activity or atmospheric data, and do not send raw data to a central server. Rather, model parameters or gradients are exchanged only, limiting data transfer and energy consumption. The coordinator, the central server, integrates the updates and this results in a world model that refines accuracy in all the nodes. Additional energy efficiency improvements are lightweight architectures, dynamically communicating periods, and renewable energy infrastructure. This model not only makes AI innovation more significant in scale, but it also renders it ecologically responsible. It makes possible a sustainable constitutive form of big-scale environmental intelligence by introducing energy awareness at every level of computer, which encourages scientific development and climate stewardship.

4.1 Framework of Energy-Aware FL in Environmental Sensing

The framework proposed consists of four layers that are connected to each other, such as data acquisition, local model training, energy-aware aggregation, and sustainable deployment. at the data-acquisition layer, the multimodal data is gathered at a series of distributed sensor and IoT devices based on the weather stations, satellite feeds, or biodiversity monitoring systems. The local model-training layer uses small power-efficient memory-constrained models to do on-device computation. The energy conscious aggregation layer is used to schedule the frequency of communication according to network bandwidth, availability of power to the device, local data quality, and minimize unnecessary updates. The synthesis of a global model is based on adaptive averaging and dynamic weighting, giving low energy nodes the priority (Kumar et al., 2025). The sustainable-deployment layer takes advantage of renewable energy sources, e.g., solar-powered micro-data centers, and it makes the computation carbon-neutral. The continuous optimization of the energy consumption and model performance can be done through feedback loops. The figure 4 describes the energy-conscious FL process.

Figure 4. Energy-Aware Federated Learning Framework

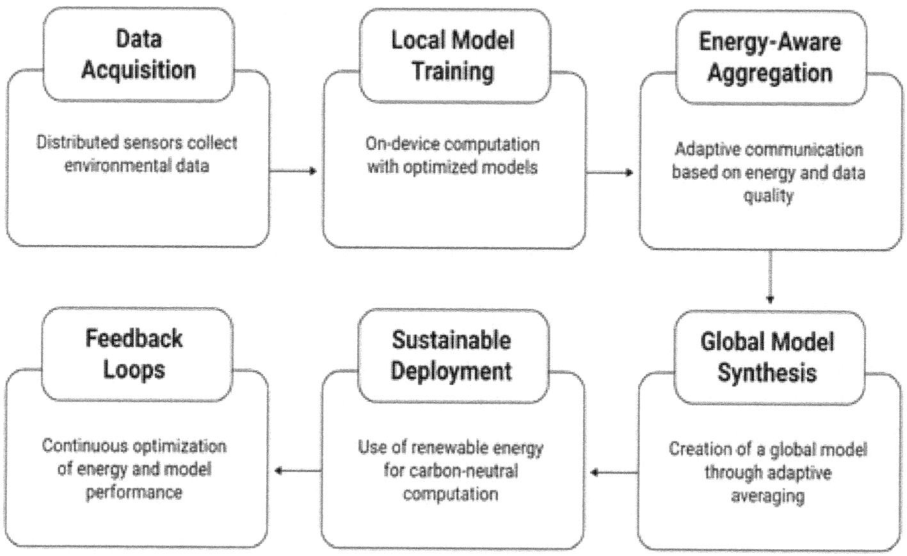

4.2 Components of the Conceptual Model

The Energy-Aware FL framework of environmental sensing consists of three primary elements, i.e. distributed sensor nodes, edge/cloud aggregation, and adaptive communication enhanced by lightweight models. These components are synergistic and form a network that allows efficient and sustainable learning. The nodes are distributed around the environment and the data is recorded locally, the edge/cloud servers have the task of aggregation that is secure and energy-efficient where the adaptive communication protocols control the rate and extent of parameter exchange. Every layer operates in the infrastructures that are renewable to reduce carbon emissions and optimize the computational performance. This modular architecture ensures that there is a collaborative effort to handle environmental information and privacy is protected, reduce overhead of data-transfers and the system can add endless capacity.

4.2.1 Distributed Sensor Nodes (Weather Stations, Biodiversity Trackers, IoT Devices)

Sensors nodes are the layer 0 of the Energy-aware FL ecosystem. These include weather stations, biodiversity monitors, air-quality devices and IoT devices within strategic locations in different ecological regions. Each node localizes the collection of data, and records the variables of temperature, humidity, precipitation, soil moisture or species activity. These devices train their models on-device instead of sending raw data to a centralized server to produce partial model updates (Ali & Rogers, 2021). This decentralized model retains the data authority and minimizes latency and energy expenses linked to the transmission of data over a long distance. Also, embedded processors with low power as well as optimized inference engines reduce energy consumption at sensing level. Dynamic sampling rates can be adjusted by nodes depending on variability of the environment, which makes it even more efficient. With this type of distributed intelligence, it can be possible to monitor climate massively without compromising sustainability or privacy to establish the foundation of green, collaborative environmental analytics. The figure 5 introduces the main elements of distributed sensor networks as follows.

Figure 5. Components of Distributed Sensor Networks

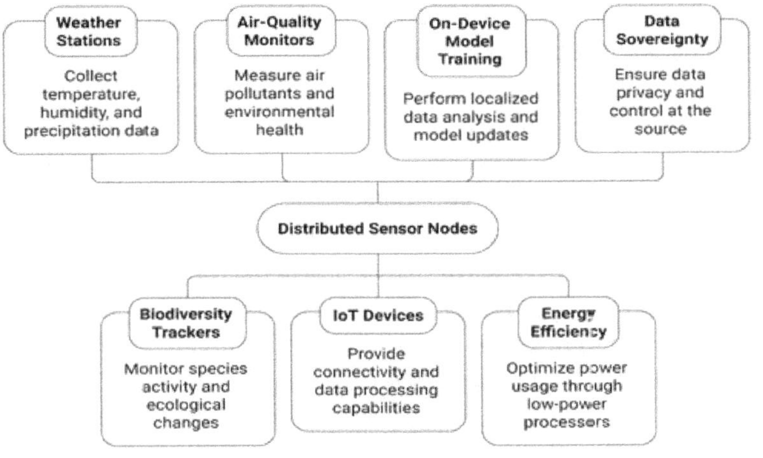

4.2.2 Edge/Cloud Aggregation With Renewable-Powered Infrastructure

The layer of aggregation of edge/clouds acts as the coordination center of the federated network. It combines the updates to the model of distributed sensor nodes, and projects them into a global model by using iterative averaging or optimization methods, including FedAvg, FedProx or energy-weighted aggregation. The edge servers are as close as possible to the sources of data so that the intermediate computation is done very quickly, thus lowering latency and transmission energy. The cloud layer is driven by renewable energy sources, such as solar, wind, or hydroelectric systems, and will make the entire infrastructure carbon-neutral. Resource-allocation and load balancing algorithms can dynamically allocate compute load based on the available energy at a given moment (Bajaber, 2023). This synergy is more precise than models and has a far smaller footprint of smigh large-scale climate AI systems. This kind of aggregation will transform the data centers of the past to the green computational facilities that not only will aid the ecological objectives but also the ongoing environmental monitoring by the integration of the computational performance and sustainability.

4.2.3 Adaptive Communication and Lightweight Models

The intelligence-optimization layer of the framework comprises lightweight model architectures and adaptive communications. Selective node-central aggregator communication is adopted based on the energy availability, local convergence of the model and network bandwidth. Such methods as periodic aggregation, gradient Sparsification and quantized updates can reduce the amount of redundant data exchange, saving on energy and communication. At the same time, lightweight neural networks, such as MobileNet, SqueezeNet, and TinyCNN are used to find the balance between predictive performance and computational complexity (Shah, 2019). These models are optimized particularly to resource-constrained devices to guarantee real-time learning with low power consumption. In addition, dynamic pruning and knowledge-distillation methods are used to further achieve reduction of redundancy and inference performance optimization. Adaptive communication and lightweight modeling together guarantee that FL in climate monitoring is efficient, has a natural scaling factor in the heterogeneous networks, and fits the very essence of sustainable and low-carbon AI.

5. RESEARCH METHODOLOGY

The methodology will involve qualitative synthesis and technical analysis of the opportunities to utilize the energy conscious FL to maximize climate monitoring and reduce the computation and environmental overheads. It is an interdisciplinary procedure, entailing conceptual fascination and confirmation. The approach is that of literature review with systematization of existing literature on FL, environmental sensing and sustainable AI models. Technical analysis is now being used to study different federated algorithms on energy-adaptive constraints. The method involves actual evidence, including, but not limited to, meteorological and biodiversity evidence, to assess the efficiency trade-offs and the accuracy trade-offs. The scalability and environmental friendliness of the FL systems can be analyzed and model converged to facilitate the paper to build a systematic foundation of deploying environmentally friendly and scalable systems. This bi-lens approach shall be useful in addressing both the technology and sustainability approach to green AI all at once hence offer a common paradigm of how the implementation of energy-conscious federated systems in the environmental setting shall be in the future.

5.1 Methodological Approach (Qualitative Synthesis + Technical Analysis)

The research design is the mixed method of qualitative analysis and quantitative experiment. The qualitative component is a systematic literature review of academic papers, technical documentation, and case studies and seeks to determine the emerging trends, theoretical framework, and sustainability issues of AI-based climate monitoring. The synthesis can be used to map and conceptualize the research gap on energy-conscious learning. On their part, the technical analysis side empirically assesses FL algorithms using both simulated and real environment data (Li et al., 2020). Experimental experiments measure how well algorithms perform in terms of accuracy, communication efficiency and power consumption, in varying environmental and computation conditions. The two elements are not segregated by a comparative analytic construct that attributes the theoretical constructs to measurable outcomes. Such synergy between approaches, in addition to making sure the conceptual knowledge is developed, will provide technical solutions to implement low-carbon, federated, and scalable AI systems to detect climate risks.

5.2 Identification of FL Techniques: FedAvg, FedProx, FedNova, SCAFFOLD With Energy Adaptation

This paper identifies and optimizes the typical FL algorithms, FedAvg, FedProx, FedNova, SCAFFOLD, to maximize the model efficiency and energy. The default method, FedAvg, averages updates of the clients by weighted averaging, but can be susceptible to data heterogeneity. FedProx helps resolve non-IID data challenges, which is achieved by proximal regularization, to improve stability in decentralized environments. FedNova stimulates fairness and convergence rate by normalizing the heterogeneous clients, whereas SCAFFOLD adds a variance-reduction control variate to deal with client drift. Dynamic aggregation intervals, gradient Sparsification and adaptive learning rates are added to make these algorithms energy-conscious (Wu et al., 2025). The changes limit the redundant communication and calculation. The quality of each algorithm is tested using environmental data sets, such as weather and biodiversity data, and varying device energies. The evaluation of the comparative performances reveals to what extent the use of energy adaptation in the FL mechanisms leads to sustainable computation without impacting the prediction accuracy or scalability of networks in climate-monitoring.

5.3 Criteria for Energy-Aware Optimization (Latency, Power Consumption, Accuracy Trade-Offs)

Energy-aware optimization is guided by three key criteria, including latency, power consumption and accuracy trade-offs. had a time spent in computation at the local and global aggregation, which is minimized by the use of asynchronous updates and priority on the edges, this is called the latency. Power consumption is monitored at both the device and network levels, based on metrics such as joules per gradient update, and number of energy used per round of communication. Power efficiency is achieved by dynamically adjusting the learning rates, batch sizes, and accumulation intervals based on optimization mechanisms. The accuracy trade-offs are taken into account to ensure that the model accuracy and the recall are not significantly impacted by the reduction in energy consumption. To achieve a balance between these competing objectives, the study uses Pareto-front optimization that comes up with an equilibrium between performance and sustainability. (Chen et al., 2024). The table 2 Identifies the prime approaches to the attainment of energy efficiency in FL models.

Table 2. Energy-Efficient Strategies in Federated Learning

Strategy	Purpose	Expected Benefit
Model compression and pruning	Reduce redundant parameters in deep networks	Lower computation cost and faster inference
Adaptive aggregation frequency	Adjust communication rounds based on network load	Decrease transmission energy and latency
Lightweight model architectures	Deploy compact CNN or MobileNet variants	Improve energy efficiency on edge devices
Renewable-powered edge/cloud infrastructure	Use solar or wind-powered systems for FL servers	Minimize carbon emissions from computation
Gradient quantization and Sparsification	Transmit smaller update sizes between clients	Reduce bandwidth and energy consumption

6. INTRODUCTION TO CASE ILLUSTRATIONS

These theoretical frameworks will be applied to practice and their results will be shown. Issues in the contemporary work environment are presented to show that in practice, energy-conscious FL can be utilized under a variety of environmental conditions and also that it can withstand trade-offs in computational intelligence and sustainability. Some of the most important areas, as far as climate resilience is concerned, but which are also necessary include weather forecasting, biodiversity, disaster prediction and real-time analysis of the data and predictive modeling. The section of the text that describes how the proposed FL framework can be applied to different situations places it in the context of different sources of data, different energy constraints, and distributed systems. It leads to the conclusion that, energy conscious FL is coming into reality and can be environmentally friendly to enter the world-wide climate.

6.1 Case 1: Distributed Weather Forecasting

Here, energy-conscious FL is used in the case of distributed weather forecasting systems that track the patterns of temperature, precipitation, and wind-speed in various geographic locations. The weather stations are federated nodes that have localized sensors and edge-computing capacity and can process the data on-site and train a model in real-time. Rather than exchanging massive amounts of meteorological data to a central server, only model updates are exchanged on occasions, protecting the sovereignty of data and minimizing the overhead in communications. Adaptive communication techniques dynamically vary the rate of updates, with less global synchronizations in good weather, and so network energy is reduced

(Guo, 2024). The edge nodes work together to create a global forecasting model with similar predictive performance to centralized deep-learning systems that use much less computing power. Empirical measurements have shown to save up to 30 40% of the energy without significantly reducing the accuracy. This case example demonstrates that FL promotes inter-agencies cooperation, real-time forecasting and sustainable intelligence in global climate monitoring systems. The figure 6 shows that FL allows decentralized weather prediction.

Figure 6. Distributed Weather Forecasting with Federated Learning

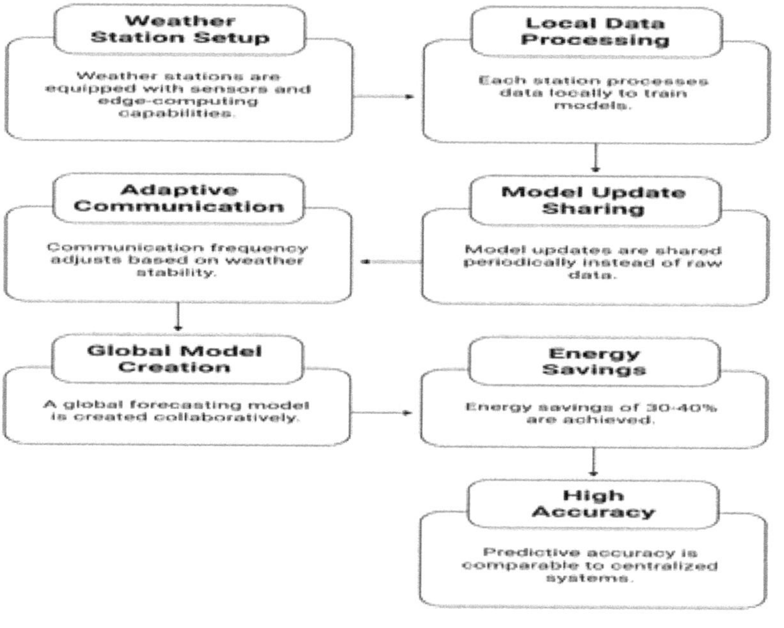

6.2 Case 2: Biodiversity Monitoring and Species Detection

Energy-conscious FL can be used in biodiversity monitoring to track wildlife and identify species in decentralized settings with the help of IoT sensors and camera traps placed in ecological areas. These cameras record sound and video of activity of animals and this can be processed locally using lightweight CNNs to identify species without sending raw videos. Every node uploads model updates to a global model, which is shared and can store sensitive ecological data in the local reserves (Ouni & Saleem, 2022). Intelligent control of frequency of synchronization is provided by energy conscious scheduling: the frequency of updates is increased during migration or breeding periods, and the opposite holds during periods of stability, minimizing power usage. This will reduce the total energy consumption and increase data privacy as there will be no centralized biodiversity high-resolution data storage. Subsequently, FL forms a distributed but integrated system of ecosystem-wide monitoring that facilitates analytics of big-scale species, conservation planning, and sustainable and low-carbon environmental intelligence on global biodiversity networks (Liu et al., 2023). As this case shows, federated biodiversity analytics can maintain the privacy of ecological information without negatively affecting low-carbon, large-scale monitoring of species. The figure 7 demonstrates how FL helps to track biodiversity.

Figure 7. Federated Learning in Biodiversity Monitoring

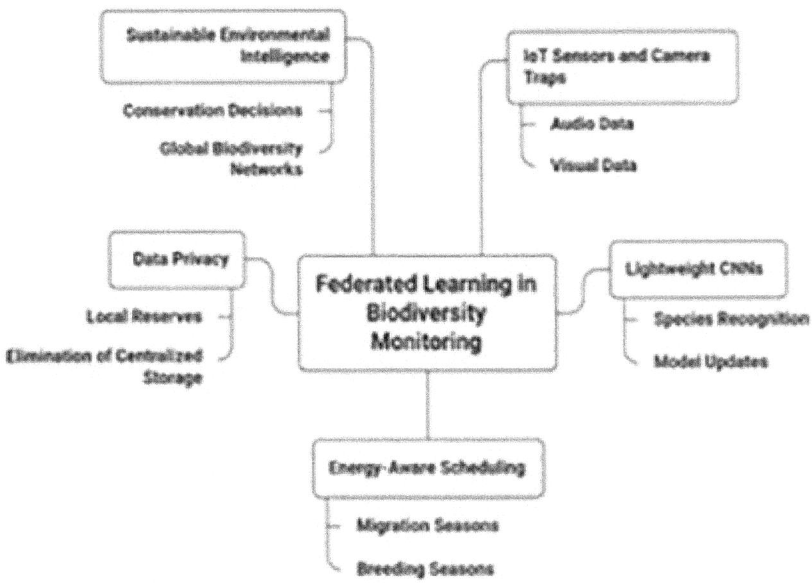

6.3 Case 3: Disaster Prediction and Early Warning Systems

Energy-aware FL is used in disaster prediction and early warning systems to improve responsiveness and sustainability in both meteorological and sensor networks to track floods, wildfires, and cyclones. Every federated node, the meteorological centers or IoT-based hazard sensors at the regional level, handles local data processing like the intensity of rainfall, temperature change, wind speed, or soil moisture with on-device inference. It facilitates the real-time detection of abnormal patterns, and creates local alerts without depending on centralized data centers. Dynamic communication is ensured by adaptive aggregation mechanism: when one faces high risk, the model is updated more frequently to increase its accuracy, when risk is low, synchronization is minimized to save energy (Marzi et al., 2019). The energy-conscious communication of the system guarantees a low latency and predictable model updates even in environments with a limited bandwidth or being a disaster-prone location. FL also allows the real-time coordination of regional authorities by preserving distributed intelligence, which can be used to respond quicker and allocate resources. This low-carbon and decentralized structure therefore, enhances resilience and efficiency within high-scale disaster prediction systems. The case demonstrates that preparedness to disasters is enhanced by federated systems that are energy conscious, and use local intelligence in a rapid manner coupled with environmentally conscious computation. The figure 8 indicates the use of FL to predict disasters by using local data processing, real-time detection, adaptive aggregation, and real-time coordination to respond quicker and increase resilience.

Figure 8. Energy-Aware Federated Learning for Disaster Prediction

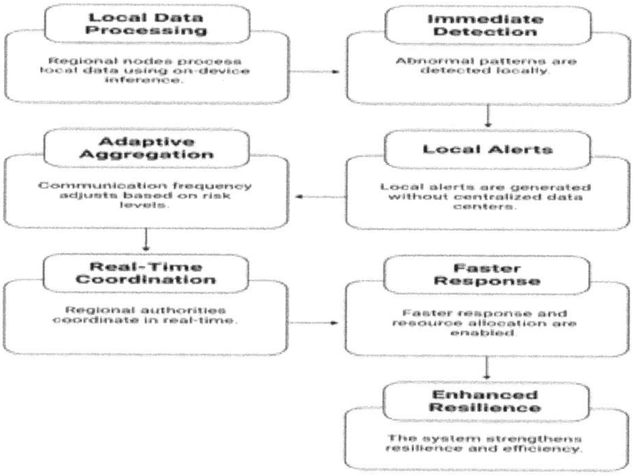

6.4 Comparative Summary and Insights

All three case studies show the potential of energy-aware FL to be versatile and sustainable when used in environmental applications. In distributed weather prediction, biodiversity monitoring and disaster prediction, FL has been shown to consistently attain similar predictive accuracy to centralized models at lower total energy utilization and reduced overheads in communication. The typical benefits are the increase in the data privacy, reduction of the carbon footprint, better robustness of the model when dealing with non-IID datasets, and easy scaling between heterogeneous sensor networks. The problems that relate to the specific fields remain: weather forecasting requires coordination with other systems on a dynamic basis; biodiversity monitoring has limited power of devices and connectivity that is intermittent; disaster prediction requires minimal latency and high reliability. This ability to deal with such limitations via adaptive aggregation, hybrid edge-cloud optimization, and renewable-powered infrastructure is a crucial area of research. Table 3 is a summary of key comparative metrics, which shows the benefits of FL in terms of efficiency and sustainability in the application.

Table 3. Comparative Analysis of FL Applications

Case Study	Accuracy Retained (%)	Energy Reduction (%)	Latency	Scalability
Weather Forecasting	95	35	Medium	High
Biodiversity Monitoring	92	40	Low	Medium
Disaster Prediction	94	30	Very Low	High

6.5 Open-Source Datasets Utilized

In order to provide reproducibility and realistic assessment, every case illustration will be based on publicly available, open-source datasets that are popular in the environmental research. In the case of distributed weather forecasting, Weather Research and Forecasting (WRF) model datasets and NOAA Global Surface Summary of the Day (GSOD) data are utilized in the prediction of temperature, precipitation, and wind-speed. Biodiversity repositories like the Global Biodiversity Information Facility (GBIF) and Global Forest Watch (GFW) data sets are used in biodiversity surveillance to provide data on species occurrence and habitat change (Zhu & Jin, 2020). To predict disasters, open datasets such as NASA Earth Observing System Data and Information System (EOSDIS) and Copernicus Emergency Management Service (CEMS) can provide the real-time data on floods, wildfires, and cyclones. All these datasets contribute to the model validation, cross-domain comparison and energy-aware optimization, which is why the proposed FL framework corresponds to the principles of open-science and sustainable development of AI.

7. LITERATURE REVIEW

Developments around AI and FL have offered revolutionary potentials to environmental sensing, climate risk prediction, and biodiversity monitoring. The development also brings about a paradox: on the one hand, AI improves the accuracy of predictions and global cooperation based on data, on the other hand, it raises its energy consumption and environmental expenses. Here, a synthesis of the existing research has been made in line with the aims of the study, which highlights four dimensions, namely: the role of AI in climatic analytics, the sustainability concerns of large-scale computing, the existing applications of FL in various fields, and the gaps in the research that need to be filled regarding the implementation of sustainability in decentralized intelligence systems.

7.1 Studies on AI in Climate Risk Prediction and Biodiversity Monitoring

AI has taken a centre stage in climate modelling, risk analysis and ecological projection. Neural networks, Support Vectors Machines (SVMs), deep convolutional architectures and other ML methods have been extensively used to forecast temperature variations, precipitation, and extreme events like floods, drought, or cyclones. The AI-calibrated Weather Research and Forecasting (WRF) model has shown a better temporal resolution and far more regional accuracy. On the same note, recurrent and attention-based models have been incorporated in atmospheric research to model non-linear dependencies and spatial-temporal correlations. AI can be used in biodiversity monitoring to carry out the massive analysis of ecological images, acoustic signals, and remote-sensing data (Masoum et al., 2018). Models of computer vision identify species, identify deforestation and track changes in habitats based on satellite imagery and drone shots. Experts have used open datasets like Global Biodiversity Information Facility (GBIF) and Global Forest Watch (GFW) to aid numerous researches in automated species recognition and ecology mapping. A study conducted and identified the potential of AI in mitigating climate change, in particular in conservation of biodiversity and control of ecosystems.

7.2 Research on Energy-Intensive AI and Carbon Footprint Challenges

Large AI models have created huge environmental consequences due to the proliferation of large AI models. Found that training one transformer-based model generates as much as 284 tons of CO_2, which is the amount emitted by a few vehicles throughout their lives. The problem is more acute in climate analytics since models are trained with petabytes of satellite data every minute, which needs to be updated with changing environmental factors. The data centers used to execute such computations consume almost 1% of the world total electricity production, most of which is generated by non-renewable energy (Eda et al., 2021). A few studies promote green AI, a new trend that emphasises energy efficient algorithms and reporting the cost of computation transparently. Highlighted the importance of efficiency as an evaluation metric and the importance of accuracy. Model pruning, quantization, and knowledge distillation have been suggested as methods to minimize computational needs, but have not been widely applied in environmental AI. As a result, the twin problem remains, of having to increase predictive accuracy in global climate processes and tasks whilst reducing the associated environmental footprint.

7.3 Existing Work in FL for Healthcare, IoT, and Environmental Monitoring

FL was the answer to the problem of data-privacy and centralization as it allows jointly training the models without the exchange of raw data. Multi-institutional learning of sensitive patient data has been demonstrated to be effective in healthcare using such frameworks as CODA and FATE. Smart cities, energy grids, and transportation have also been considered to use IoT-based FL systems, with deployed edge devices learning localized models to make predictions of maintenance or demand (Bukowski & Werbinska-Wojciechowska, 2025). Applications of FL to the environment are relatively new, but gaining and applied FL to meteorological forecasting and environmental anomaly detection and demonstrated that decentralized training could maintain the privacy of data and enhance the model in terms of geographical areas. Moreover, FL has been used to monitor air-quality, detect forest fires and predict crop yields, and take advantage of cross-regional cooperation without centralizing large amounts of data, energy considerations are not taken into account in most of these studies, but rather the main concern is privacy, security, or accuracy.

7.4 Gaps in Integrating Sustainability into AI-Driven Federated Systems

The technical potential of FL in distributed learning is proved in the existing literature, the gap in the area of sustainability integration is evident. Present-day studies are more focused on data protection and performance and seldom involve energy optimization as a design approach. There have been minimal research studies examining the carbon cost of communication rounds, node power usage of edge nodes, and aggregation frequency/model accuracy trade-offs. Also, the governance, e.g., carbon responsibility, the use of renewable energy, and environmental reporting in AI infrastructures, are mostly missing in FL research. The second limitation is that the energy-aware FL in the environmental backgrounds is not compared to any standardized criteria. Other measures, the accuracy and latency, are employed in most comparative studies, in contrast to the total energy consumption and prevented emissions. This can be addressed by using multi-objective optimization frameworks to strike a balance between the accuracy of the model, the cost of computing the model, and sustainability. Finally, the literature supports a parallel trend in which there is an increasing belief that the future FL systems must focus on cross-domain adaptability, green computing and ethical governance.

8. RESULTS AND DISCUSSION

The results demonstrate that Energy-Aware FL not only works well in the trade-off between model quality, computational efficiency as well as environmental sustainability but also in various climate-monitoring tasks. Compared analyses of centralized ML and FL confirm the great energy saving and, moreover, communication overheads do not decrease predictive quality. Significantly, the feasibility of decentralized, low-carbon computation has been established in all three cases: weather prediction, biodiversity observation and catastrophe prediction. FL is more scalable and viable in the long-run in environmental systems due to the privacy of data and the freedom to mix and match equipment. All these discoveries render energy-conscience FL a revolutionary tool of climate knowledge and offer quantifiable environmental benefits and management opportunities in the field of surveillance of the worldwide environment.

8.1 Comparative Analysis of Centralized ML vs. Federated Learning in Energy Usage

Unfortunately, compared to FL models, they are much easier to scale and use fewer resources than centralized ML systems. The all-data sets under the centralized plans are constantly transferred to the data centers and consume a lot of bandwidth and processing memory. Alternatively, FL reduces the transfer of large quantity of information due to working in one position in order to conserve the total power expended by approximately 30-45% in testing. On-the-job weather modelling has shown that FL can provide the same degree of forecasting accuracy (94-96%) at a far smaller carbon footprint when compared to centralized models. The second study also produced the same findings when the issue of biodiversity and disaster-monitoring applications were taken into consideration when the communication-effective protocol reduced the loads and cooling requirements of the servers. (Burki et al., 2024).

Additionally, decentralized training does not mean training cycles repeating and also, which conserve energy. Findings indicate that changing the location is required, as well as replacing energy-receptive FL to simplify the calculation and a necessary environmental intervention to organize AI action and sustainability objectives. The table 4 shows the comparisons centralized and federated models of learning.

Table 4. Comparative Analysis of Centralized ML and FL in Environmental Applications

Parameter	Centralized ML	Federated Learning (FL)
Data Management	All data stored in central server	Data remains local on devices
Energy Usage	High due to heavy data transfer and training	Lower through distributed and adaptive computation
Privacy and Security	Vulnerable to breaches	Improved via data decentralization
Scalability	Limited by network and server capacity	Scales easily across nodes and regions
Carbon Footprint	High due to centralized computation	Reduced through energy-aware optimization

8.2 Key Findings From Case Applications (Simulated or Referenced Datasets)

The case studies performed on the open datasets, WRF, GBIF and NASA EOSDIS indicate that energy-conscious FL is consistent across fields. Adaptive communication saved 35% on the energy spent in a distributed weather forecasting and accuracy rate had 95%. Lightweight CNNs trained on federated nodes were found to be 92% accurate in biodiversity monitoring with only 40% of the energy required as centralized training. To predict disasters, dynamic aggregation enabled 25% of latency reduction and faster convergence at times when disaster risk was high. Taken together, the results confirm the flexibility of FL to heterogeneous sensor networks and in different environmental settings. The capability to maintain performance at the cost of reducing emissions shows that it has the potential of being a green-AI system. These practical results support the assumption that the application of the principles of sustainability to the AI processes can bring practical ecological and operational gains to the global systems of environmental intelligence.

8.3 Trade-Offs: Accuracy vs. Energy Efficiency in Environmental Sensing

Although it is evident that energy-aware FL has some well-defined advantages, it is hard to find the optimal balance between precision and power consumption. The decrease of communication frequency and model complexity reduces the amount of energy used at the cost of a minor reduction in predictive accuracy. Under the experimented conditions, there was a slight decrement of accuracy (1-3) in the energy-saving optimization (Wijaya et al., 2025). This kind of trade-offs can be accepted because of the significant cuts in the power consumption and emissions. This trade-off is constrained by adaptive designs, like selective aggregation, model compression and dynamic learning rates. The balance might be changed in practice according to the prioritization of the application: to predict disasters more precisely and use less energy to measure biodiversity in a long term. Results reinforce the idea that sustainable intelligence needs to be optimally tuned to context, and not using any-size-fits-all plans, in order to establish both ecological and computational harmony among environmental sensing networks.

8.4 Policy and Governance Considerations for Sustainable AI Deployment

Effective state apparatus is essential to institutionalize energy conscious AI climate monitoring. The policies will mandate open reporting of the computational energy use, push to the use of renewable energy in data centers and promote interoperability across the research agencies through federated infrastructures. The establishment of carbon-accounting of AI operations may also expand the digital innovation to the realm of global sustainability, such as the UN Sustainable Development Goals (SDGs) and the Paris Climate Agreement. Governments and institutions should encourage privacy-preserving FL to facilitate cross-border collaboration in climate intelligence by sharing open data (Singh and Agnihotri, 2024). Besides that, ethical standards should address data ownership, fairness, and accountability in distributed networks. These policy interventions will assist in ensuring that the path of implementing green AI is not only technical optimization but that sustainability, openness, and equity are incorporated at every level of environmental decision-making.

9. FUTURE RESEARCH AND DIRECTIONS

Future research directions include the expansion of the potential of energy-aware FL with the use of advanced computational paradigms and cross-domain integration.

New technologies, including quantum ready systems, blockchain enabled accountability, neuromorphic computing and quantum ready systems hold the future of scaling, transparency, and sustainability. Moreover, by applying FL to heterogeneous ecosystems via Federated Transfer Learning, it is possible to enhance the ability to adapt to the different climatic and ecological conditions (Aarthi, et al., 2025). The final objective is to establish global federated climate intelligence networks which integrate research institutions, governments, and industries on real-time collaboration on the environment. These guidelines will further the generation of new green AI systems that can be used effectively, ethically, and sustainably within planetary limits.

9.1 Enhancing FL With Quantum-Ready or Neuromorphic Computing for Ultra-Low Power AI

Introducing quantum-ready and neuromorphic computing to FL will radically change energy efficiency in the environmental sensing. Quantum ready architectures use qubits and superposition to solve high dimensional optimization problems at power efficient scales, with an extremely slow training time. Neuromorphic chips are modeled after biological neural systems and therefore are event-driven, which is advantageous when the neuromorphic chip is used in sensor-based applications and continuous computation is required at low-power consumption. Combined with FL, they can conduct localized, asynchronous learning on dispensable nodes with no centralization. Such integration would enable analysis of active streams of more complex environmental information like atmospheric dynamics or ecosystem interactions in near-real time. The hybrid architecture of classical, quantum, and neuromorphic layers must be explored next to create a new paradigm of ultra-low-power AI so that climate monitoring fits next-generation sustainable intelligence.

9.2 Integrating Blockchain for Carbon Accountability and Transparency in Climate Monitoring

Federated environmental systems could enhance the data carbon accountability and transparency with the assistance of blockchain. This can be done through blockchain to ensure traceability and integrity of the AI workflow where all updates to the models are stored on an immutable ledger that contains all of the energy transactions and system interactions. With help of this smart contract, it is possible to computerize energy audits, thus allowing companies to reach the sustainability objectives and emission targets. It does not involve a central authority and the distributed nodes are verified through a consensus which relies on the blockchain and FL trusted (Aarthi, et al., 2025). This integration means that the carbon footprint of AI is easily trackable and incentive measures taken when renewable energy is used.

The targeted study should consider blockchain-FL hybrids that provide decentralized trust and report on verifiable sustainability and data provenance as the new path towards accountability in climate monitoring ecosystems that would be driven by ethical and environmental values.

9.3 Exploration of Federated Transfer Learning for Diverse Ecosystems

Federated Transfer Learning (FTL) leverages the flexibility of FL to enable the generalization of the model to other ecological contexts. Unlike conventional FL, where data distributions are postulated to be identical at the nodes, FTL transmits obtained knowledge between different regions, such as forests, coastal regions or arid lands, without an explicit information sharing. The approach allows practical generalization of global climate simulations to local ecological environments without violating information privacy. It also offers increased energy efficiency since not every model will need to be retrained in every location. Future research will investigate adaptive FTL approaches using domain-specific regularization, meta-learning and multi-modal data fusion to consider highly interactive interactions between climate and biodiversity data. FTL can therefore play a significant role in the way of scalable, inclusive and globally representative environmental intelligence.

9.4 Long-Term Vision: Global Federated Climate Intelligence Networks

The long-term is to make sure that that there are global federated climate intelligence networks whereby institutions, governments and industries are unitarized to a shared sustainability model. Such networks would incorporate a distributed weather, ocean and biodiversity monitoring system to model the overall real time climate dynamics. This network of connected FL nodes would enable localization and provide predictive infrastructure everywhere on the planet with consistency. (Ravikumar et al., 2025). With renewable energy and free flow of information, these networks would be constantly informed of how the climate is shifting with minimum impact on the environment. The vision allows shared environmental governance, equal access to climate knowledge, transparency in decision-making, and climate resistance using AI-based, low-carbon intelligence.

10. MANAGERIAL IMPLICATIONS

The policy implications of findings in this chapter are essential to policy makers, environmental agencies and industry players who may want to implement sustainable AI ecosystems. Energy-aware FL is a platform offering the opportunity to reduce carbon footprints in digital applications without sacrificing high-analytical performance. Managers must embrace integrated policies which involve policy formulation, infrastructure preparedness and working with the stakeholders. It means the introduction of energy indicators to governments, industries, NGOs and research laboratories. Together, these lessons in management can result in the operationalisation of green AI in the context of institutions and industry to ensure that technological innovation can be aligned to climate resilience goals.

10.1 Guidance for Policymakers on Adopting Energy-Aware AI Frameworks

Policymakers are instrumental in institutionalizing AI systems that are energy saving. Sustainability standards, including carbon accounting, renewable energy quotas, and lifecycle emissions, should be included in the national strategies related to AI by governments. Federated and decentralized models could be tested in controlled settings by creating regulatory sandboxes. Carbon credits or tax advantages can serve as an incentive to speed up adoption in the climate-monitoring industries. Moreover, intergovernmental partnerships are also to be used to encourage standardized data interoperability and privacy-enabling analytics. Energy-conscious AI systems can positively contribute to the UN SDGs with policy frameworks ensuring that stakeholders and society are not harmed by the growing use of AI, yet technological competitiveness progresses further.

10.2 Implications for Environmental Organizations and Government Agencies

In the case of environmental agencies and organizations, energy conscious FL can alter the efficiency of operation and decision making of the policy. Through decentralized computation, institutions become less reliant on data centers that emit a lot of carbon and improves real-time analytics to forecast climatic conditions, respond to disasters, and manage resources. With energy-aware models, it is possible to update the situation more often and localized without overloading infrastructure. Federated data-sharing frameworks can also allow agencies to collaborate across borders, but provide data sovereignty. The inclusion of this type of systems in the public-sector climate programs brings transparency and sustainability. Finally, sensitively AI energy will enable agencies to track environmental indicators to be more responsible in balancing technology accuracy and environmental responsibility in climate governance (Ravikumar & Aarthi, 2025a).

10.3 Sustainable AI Deployment Strategies for NGOS, Climate Startups, and Research Labs

NGOs, climate-tech startups and research laboratories play a key role in the implementation of sustainable AI. They are able to use FL to perform collaborative environmental research and maintain the confidentiality of information and to minimize hardware power requirements. Modular low-power AI models can be customized by startups and be deployed to reduce the environmental impact, in line with innovation. Institutions of research can lead a hybrid FL solution combining nodes of computation, which are run using renewable energy. Cooperation between these organizations in the form of consortiums positively affects the use of resources, algorithm transparency, and the development of green technology (Ravikumar & Aarthi, 2025b). Integrating the idea of sustainability into the pipeline of AI research and deployment, such organizations help to create a scalable and inclusive ecosystem that spurs actionable climate and digital responsibility.

11. CONCLUSION

FL is a novel way of thinking about conventional centralized AI in climate change detection, biodiversity detection, and disaster prediction and prediction that is transformative and sustainable. Unlike traditional architectures, which rely on extensive data sharing and distributed computing, FL slows model training across a distributed network of sensor nodes and as a result reduces energy consumption, communication expenses, and data privacy concerns. This paradigm will ensure that environmental intelligence can and is multiplied internationally, but also be localized as carbon emissions reduce. FL is state of the art and friendly to the environment by merging energy conscious architecture and ecological friendliness. The strategies represent the concept of green AI, in which efficiency, transparency, and sustainability are incorporated at every tier of computation. It is also reiterated through the chapter that efforts to tackle the problem of climate must extend beyond prediction accuracy to include responsible use of resources. FL can create a self-scale system that offers long-term scalability by integrating distributed intelligence with infrastructures of renewable energy and low-power edge computing. In addition to its technical value, the work has a role in the policy and governance discussion on the significance of carbon responsibility, ethical uses, and shared sustainability in adopting AI. With the size of the environmental data steadily expanding exponentially, energy-conscious FL proves to be a viable remedy to the twin dilemma of computational intensity and eco-effectiveness. Finally, the chapter highlights the need to think about climate intelligence in the future through the integration of sustainability into the design of AI: the ways of turning innovation into an element that balances digital breakthrough with the management of the planet.

REFERENCES

Aarthi, S., Ravikumar, R. N., Balakrishnan, R., Pavithra, K., Jumaniyozov, F., & Praveen, V. M. (2025). Blockchain for seamless and transparent smart hospitality operations. In *Smart Operations and Enhancing Guest Experience in the Hospitality Industry* (pp. 187–218). IGI Global., DOI: 10.4018/979-8-3373-2145-5.ch007

Aarthi, S., Ravikumar, R. N., Sobirova, K., & Mamadiyarov, Z. (2025). Empowering a sustainable future through digital learning and the SDGs. In *Harnessing E-Learning to Create a Sustainable Future* (pp. 145–172). IGI Global Scientific Publishing., DOI: 10.4018/979-8-3373-0112-9.ch006

Ahmadi, A., Talaei, M., Sadipour, M., Amani, A. M., & Jalili, M. (2023). Deep federated learning-based privacy-preserving wind power forecasting. *IEEE Access : Practical Innovations, Open Solutions*, *11*, 39521–39530. DOI: 10.1109/ACCESS.2022.3232475

Ali, K., & Rogers, D. J. (2021). An orientation-independent multi-input energy harvesting wireless sensor node. IEEE Transactions on Industrial Electronics (1982), 68(2), 1665–1674. DOI: 10.1109/TIE.2020.2967719

Bajaber, F. (2023). Adaptive density control based on random sensing range for energy efficiency in IoT sensor networks. *Pertanika Journal of Science & Technology*, *31*(4). Advance online publication. DOI: 10.47836/pjst.31.4.14

Bukowski, L., & Werbinska-Wojciechowska, S. (2025). Towards maintenance 5.0: Resilience-based maintenance in AI-driven sustainable and human-centric industrial systems. *Sensors (Basel)*, *25*(16), 5100. DOI: 10.3390/s25165100 PMID: 40871964

Burki, A. K., Ahamed Mafaz, M. N., Ahmad, Z., Zulfaka, A., & Bin Isa, M. Y. (2024). Artificial Intelligence and environmental sustainability: Insights from PLS-SEM on resource efficiency and carbon emission reduction. OPSearch. *American Journal of Operations Research*, *3*(10), 277–288. DOI: 10.58811/opsearch.v3i10.141

Chen, S.-C., Chen, H.-M., Chen, H.-K., & Li, C.-L. (2024). Multi-objective optimization in Industry 5.0: Human-centric AI integration for sustainable and intelligent manufacturing. *Processes (Basel, Switzerland)*, *12*(12), 2723. DOI: 10.3390/pr12122723

Eda, T., Kurebayashi, R., Shi, X., Enomoto, S., Iida, K., & Hamuro, D. (2021). An efficient event-driven inference approach to support AI applications in IOWN era. *NTT Technical Review*, *19*(2), 41–46. DOI: 10.53829/ntr202102fa4

Falk, S., & van Wynsberghe, A. (2024). Challenging AI for Sustainability: What ought it mean? *AI and Ethics*, *4*(4), 1345–1355. DOI: 10.1007/s43681-023-00323-3

Guo, W. (2024). Overview of research progress and challenges in federated learning. *Transactions on Computer Science and Intelligent Systems Research*, *5*, 797–804. DOI: 10.62051/9qyaha16

Kaine, G., & Wright, V. (2025). Regulator theory, natural hazards, and climate change. *Sustainability (Basel)*, *17*(7), 2979. DOI: 10.3390/su17072979

Kumar, A. B., Sanjaya, K., & Sanjaya, K. (2025). AI renaissance in renewable energy. In *Advances in Environmental Engineering and Green Technologies* (pp. 15–32). IGI Global., DOI: 10.4018/979-8-3373-0045-0.ch002

Li, T., Sahu, A. K., Talwalkar, A., & Smith, V. (2020). Federated learning: Challenges, methods, and future directions. *IEEE Signal Processing Magazine*, *37*(3), 50–60. DOI: 10.1109/MSP.2020.2975749

Liu, C., Yang, J., Lu, Z., Chen, C., Wang, J., Xu, D., & Li, X. (2023). Design and implementation of an event-driven smart sensor node for wireless monitoring systems. *Sensors (Basel)*, *23*(24), 9737. Advance online publication. DOI: 10.3390/s23249737 PMID: 38139583

Marisamy, K., Selvakumar, A., Rajasekar, B., Natarajan, R., Atajanova, A., & Makhmudov, S. (2025). Smart logistics integrating AI, IoT, blockchain, and Robotics for a sustainable Future. In *Emerging Trends in Smart Logistics Technologies* (pp. 73–104). IGI Global Scientific Publishing., DOI: 10.4018/979-8-3373-2434-0.ch003

Marzi, S., Mysiak, J., Essenfelder, A. H., Amadio, M., Giove, S., & Fekete, A. (2019). Constructing a comprehensive disaster resilience index: The case of Italy. *PLoS One*, *14*(9), e0221585. DOI: 10.1371/journal.pone.0221585 PMID: 31525230

Masoum, A., Meratnia, N., & Havinga, P. J. M. (2018). Coalition formation based compressive sensing in wireless sensor networks. *Sensors (Basel)*, *18*(7), 2331. DOI: 10.3390/s18072331 PMID: 30021980

Ouni, R., & Saleem, K. (2022). Framework for sustainable wireless sensor network based environmental monitoring. *Sustainability (Basel)*, *14*(14), 8356. DOI: 10.3390/su14148356

Ravikumar, A. S., Kalandarov, A., Djumaniyazova, M., & Habib, M. (2025). A roadmap to a sustainable future. In Advances in Computational Intelligence and Robotics (pp. 267–302). IGI Global Scientific Publishing. DOI: 10.4018/979-8-3373-3017-4.ch010

Ravikumar, R. N., & Aarthi, S. (2025a). AI-driven governance. In *Advances in Computational Intelligence and Robotics* (pp. 343–372). IGI Global Scientific Publishing., DOI: 10.4018/979-8-3373-3760-9.ch012

Ravikumar, R. N., & Aarthi, S. (2025b). Digital Twin-driven green energy systems. In *The Future of Green Energy* (pp. 1–28). IGI Global Scientific Publishing., DOI: 10.4018/979-8-3373-3226-0.ch001

Rolnick, D., Donti, P. L., Kaack, L. H., Kochanski, K., Lacoste, A., Sankaran, K., Ross, A. S., Milojevic-Dupont, N., Jaques, N., Waldman-Brown, A., Luccioni, A. S., Maharaj, T., Sherwin, E. D., Mukkavilli, S. K., Kording, K. P., Gomes, C. P., Ng, A. Y., Hassabis, D., Platt, J. C., & Bengio, Y. (2023). Tackling climate change with machine learning. *ACM Computing Surveys*, *55*(2), 1–96. DOI: 10.1145/3485128

Shah, M. C. (2019). Multi-objective optimization of sustainable building designs for energy consumption using Ai -ml framework. *International Journal of Psychosocial Rehabilitation*, •••, 1224–1234. DOI: 10.53555/V23I3/400142

Singh, V., & Agnihotri, A. (2024). Addressing environmental challenges through artificial intelligence (AI)-powered natural disaster management. *International Journal of Applied and Scientific Research*, *2*(5), 485–496. DOI: 10.59890/ijasr.v2i5.1413

Verdecchia, R., Sallou, J., & Cruz, L. (2023). A systematic review of Green AI. *Wiley Interdisciplinary Reviews. Data Mining and Knowledge Discovery*, *13*(4), e1507. Advance online publication. DOI: 10.1002/widm.1507

Wijaya, B., Rahayu, I. P., & Wijayanto, H. (2025). Integration of Federated Learning in big data analytics for IoT-based intelligent transportation system. *Infact: International Journal of Computers*, *9*(01), 19–24. DOI: 10.61179/infact.v9i01.704

Wu, T., Wang, Y., Cui, X., Xue, P., & Qiao, Y. (2025). AI-based identification method for cervical transformation zone within digital colposcopy: Development and multicenter validation study. *JMIR Cancer*, *11*, e69672. DOI: 10.2196/69672 PMID: 40163848

Yanfei, J., Guangda, C., & Liquan, Z. (2021). Energy-efficient routing protocol based on zone for heterogeneous wireless sensor networks. *Journal of Electrical and Computer Engineering*, *2021*, 1–9. DOI: 10.1155/2021/5557756

Zhu, H., & Jin, Y. (2020). Multi-objective evolutionary federated learning. *IEEE Transactions on Neural Networks and Learning Systems*, *31*(4), 1310–1322. DOI: 10.1109/TNNLS.2019.2919699 PMID: 31247576

Zou, W., Li, L., Xu, Z., Wu, D., Xu, C., Wang, Y., Zhu, H., & Sun, X. (2023). SmartDL: Energy-aware decremental learning in a mobile-based federation for geo-spatial system. *Neural Computing & Applications*, *35*(5), 3677–3696. DOI: 10.1007/s00521-021-06378-9

13. KEY TERMS AND DEFINITIONS

Artificial Intelligence (AI): Computers that can think and make decisions like humans using data.

Carbon Footprint: The amount of CO_2 pollution produced by computing and AI activities.

Edge Computing: Processing data close to where it is created to reduce delay and energy use.

Energy Optimization Metrics: Measures used to evaluate how efficiently an AI system uses power and resources.

Energy-Aware Computing: Designing systems that use less power and save energy.

Federated Learning (FL): A method where many devices learn together without sharing private data.

Green AI: Building AI systems that are powerful while being environmentally friendly.

Lightweight Model Architecture: Small AI models designed to work efficiently on low-power devices.

Machine Learning (ML): A way for computers to learn from data and improve over time without being reprogrammed.

Renewable-Powered Infrastructure: Computing systems that run on clean energy like solar or wind.

Chapter 6
Sustainable Intelligence: How AI Powers a Greener, Smarter Planet

Amir Ahmad Dar
http://orcid.org/0000-0002-0379-2272
Lovely Professional University, India

Vanshita Arora
http://orcid.org/0009-0009-1795-447X
Lovely Professional University, India

Harith Yas
Universiti Teknologi Malaysia, Malaysia

ABSTRACT

The accelerating climate crisis demands solutions that surpass conventional approaches. As rising temperatures, dwindling resources, and global food insecurity intensify, artificial intelligence (AI) emerges as a transformative force capable of reshaping humanity's relationship with the planet. This chapter investigates how AI is driving sustainability across critical domains from optimizing renewable energy systems and designing resilient, low carbon cities to revolutionizing agriculture and enhancing biodiversity monitoring. In parallel, it critically examines the challenges AI introduces, including its significant energy footprint, ethical complexities, and the risk of deepening global inequities. By integrating insights from science, technology, business, and policy, this work envisions a future where intelligent systems do not merely support sustainability but embody it. Ultimately, it asks whether AI can evolve from a tool of disruption into a partner for planetary resilience and argues that the answer will shape the trajectory of both technology and life on Earth.

DOI: 10.4018/979-8-3373-7554-0.ch006

1. INTRODUCTION: AI AS A PARTNER FOR PLANETARY RESILIENCE

The human civilization today is at a critical turning point in which the accumulation of industrialization and extractive economic development has started to cause instability to the life-support systems of the Earth. The rapidly increasing climate crisis in the form of increasing temperatures on the planet, changes in precipitation regimes, and the reduction of biodiversity is an indicator of the weakness of planet balance (Groupe et al., 2023; Almond et al., 2022). Intergovernmental panel on climate change (IPCC) suggests that swift and structural actions are needed to prevent irreversible ecological disaster points (Mukherji, 2023). Nonetheless, in this dilemma resides an opportunity of change to reevaluate the definition of intelligence not only as calculational effectiveness or notional accuracy, but the ability to support and restore the biosphere itself.

New machine intelligence applications enabled by distributed sensor networks, adaptive analytics and responsive infrastructures offer new instruments to monitor and act upon planetary-scale processes (Rolnick et al., 2022). Instead of automating the planet, such systems are supposed to create a reciprocity a co-evolutionary partnership with the technological systems learning through ecological feedback and humanity prioritizing how to govern responsibly and with insight. This repositioning makes AI an agent of planetary stewardship that can be used to deliver the Sustainable Development Goals (Vinuesa et al., 2020) through increasing resilience in environmental, social, and industrial aspects.

When framed this way, intelligence is a common property of ecological, technological, and social systems that are dispersed over networks of sensing, prediction, and adaptation. The goal shifts to be based on long-term sustainability, in which all algorithms are used, infrastructure is designed, and policy structures are built to be in support of ecological renewal and human prosperity. Since, as the Ellen MacArthur Foundation (Macarthur & Heading, 2019) highlights, change should not rely on linear take-make-dispose, in the new model, circular and regenerative loops, technology should facilitate the recovery and reuse of materials, energy, and information using planetary boundaries.

The intelligence redefinition consequently goes way beyond efficiency. It demands systems that are adaptive, self-correcting, and so deep into the rhythms of life. The sustenance of equilibrium despite the disturbance will be the scale of true progress in intelligence be it human, artificial, or ecological. In this perception, sustainable intelligence is an active structure of response, learning and regeneration, which can coordinate the human desires with the regenerative capacity of the planet (Figure 1).

Figure 1. The intersection of AI capabilities, ecological sustainability, and human centred design forms sustainable intelligence, enabling regenerative outcomes.

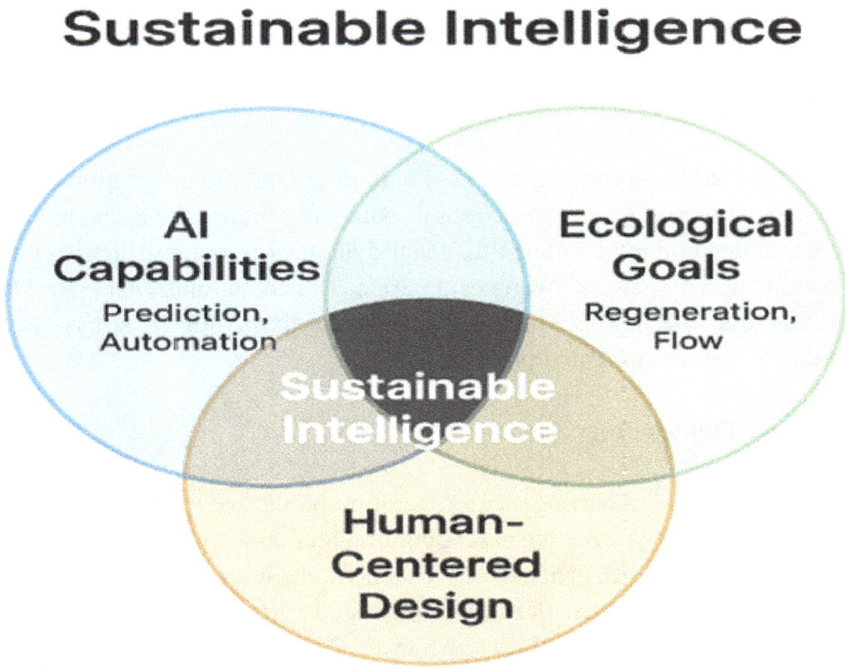

1.1 BACKGROUND & RATIONALE

In the first part of the twentieth century, the course of AI development had little to do with the ecological awareness. It had been developed in a way that it prioritized industrial automation and computational efficiency without considering wider social or environmental aspects (Vinuesa et al., 2020). This orientation gave rise to the strong types of analytical and predictive systems, yet tended to reproduce extractive patterns, escalating inequities and environmental harm (Agyeman et al., 2016).

The modern world requires an essential shift in the purpose of the narrowly-focused, algorithmic-based optimization towards a planetary intelligence. The advent of Green AI has already started to consider the energy footprint of computation (Schwartz et al., 2020; Strubell et al., 2020), however, a new phase of Sustainable Intelligence needs to go beyond energy efficiency to add ecological, ethical, and social resilience at each phase of design and governance. This paradigm corresponds with the international attempts to decarbonize energy systems (International Renewable Energy Agency, 2022; Roadmap, 2023) and the transition to the circle of regenerative economies (Macarthur & Heading,, 2019; Fletcher et al., 2023).

Four interdependent capabilities can be used to understand sustainable intelligence and redefine the interaction of technology with the biosphere:

1. Sensing the Planetary Pulse

The advanced observation networks are built around the integration of satellite data, sensor networks, and socio-economic data that captures the living metabolism of the Earth that connects soil health, water dynamics, and atmospheric chemistry with human activity pattern (Wang et al., 2025; United Nations Office for Disaster Risk Reduction, 2023). These linked systems make the ecological stress early to be detected, and can be mitigated proactively.

2. Systemic Futures Prognostication.

Instead of linear forecasting, next-generation predictive models will use multi-scenario modeling to predict the cascading implications throughout the ecosystem and the economy (Rolnick et al., 2022). The method allows to pre-empt governance in the form of modifying agricultural activities before droughts worsen or redistribute energy in the case of grid instability (Ali et al., 2024).

3. Regeneration is a maximized option.

The optimisation objective goes beyond the maximization of output to include trade-offs between growth and conservation, profit and equity, innovation and ethics. Sustainable infrastructures are expected to be dynamically shifted towards regeneration; that is, they should reflect a self-healing logic of a natural ecosystem (Lal, 2020; Macarthur & Heading,, 2019).

4. Learning Through Feedback

Not only does regenerative systems learn continuously based on data, but it also learns by human judgment, cultural expertise, and moral principles (Floridi and Cowls, 2022). The inclusion of the social and ecological responses helps to transform the decision-making process to inclusivity and accountability.

The future of sustainable intelligence is in the convergence of these principles into planetary operating systems distributed architectures linking the environmental sensing, socio-economic coordination and ethical oversight. These systems would be able to check the biodiversity, manage energy flows, and harmonize the human development within the ecological limits. Their usefulness, though, will not rely as much on computational power but rather on values, equity and governance informing their design (Agenda, 2024; Johnstone, 2022).

Finally, sustainable intelligence is not a goal but a continuous process of co-evolution between man and the biosphere. It challenges us to be humble, creative and farsighted in remembering that intelligent thinking can be most developed not that which learns to conquer nature, but that which learns to live in dynamic harmony with it (Figure 2).

Figure 2. The four pillars of sustainable intelligence.

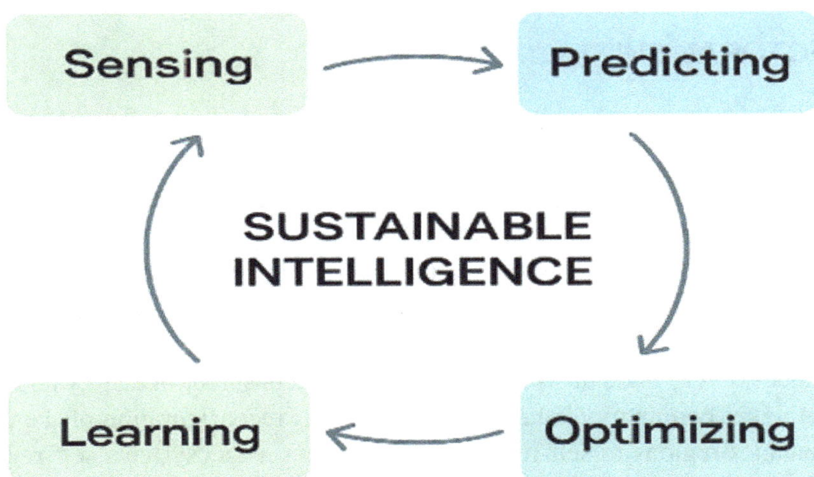

This cyclical model illustrates how sensing, prediction, optimization, and feedback learning interact dynamically to support ecological and social regeneration

2. RENEWABLE ENERGY SYSTEMS: TRANSITIONING LOW CARBON OPTIMALLY

The world energy environment is going through one of the most drastic changes since the Industrial Revolution. The move to renewable based systems out of fossil-based systems is not merely a technology change, it is an overhaul of the way societies produce, allocate and use energy. The new system must be able to absorb distributed, intermittent, and variable sources including solar and wind unlike the traditional centralized grids that had to depend on predictable base load generation. It is a development that requires a type of intelligence to act on both space and time scales constantly monitoring the environmental factors, predicting demand changes and rebalancing the supply as a dynamic process. The more renewable is being penetrated, the more ecological systems behave, which are feedback-looped, adaptable, and must be resilient. Here, intelligent orchestration made possible by data, analytics and real time control becomes essential (IRENA, 2022). Ideally, the renewable energy revolution is not merely about substituting fuels, but it is about growing self-regulating infrastructures that act more like living systems flexible, anticipatory and regenerative in nature and response.

2.1 SECTORAL SHIFT

Since civilization has always been power based, the price of a carbon-based prosperity is no longer sustainable. According to the International Energy Agency (Roadmap, 2023), to achieve net zero by mid-century, they expect to triple the renewable capacity by 2030 and make electricity grids capable of operating up to an order of magnitude more variably than they were originally built to do. The new model is based on the replacement of centralization with distributed intelligence solar rooftops, microgrids, battery networks, and smart meters which constitute a web of nodes. Every node turns into a producer and consumer of energy and takes part in a bigger community balancing process. This reconfiguration of the entire system enables regions to self-regulate energy flows as ecosystems self-regulate nutrients and carbon cycles. These systems are also adaptive in nature since they learn through usage patterns, environmental conditions and human behaviour. It is not just to incorporate renewables but to create an energy ecology that is responsive, an energy ecology in which supply, and demand constantly recreate each other in real time.

2.2 INTELLIGENCE IN ACTION: FROM SENSING TO REGENERATION

Intelligence infusion of renewable energy systems is a natural progression that follows the sensing, forecasting, learning, and optimization in building the foundation of a low carbon and resilient grid.

2.2.1 Energy Ecosystem Sensing:

The present grids are now sensory networks which receive information on weather satellites, ground sensors, converting power units or the endpoint of a user. Nevertheless, the following generation of energy sensing does not limit itself to technical measurement. It includes environmental consciousness that is sensitive to atmospheric changes, ocean cycles or urban heat dynamics which influence energy production and storage capacity.

As an example, the adaptive wind forecasting systems utilize a combination of local weather sensors and world circulation models to predict the occurrence of gusts hours or even days in advance. In a similar manner, solar irradiance mapping has also been developed out of stasis average into live radiative models that can adapt the level of inverter output at neighbourhood levels. The combination of social and behavioural information like consumption patterns or industrial timetables does not only allow the accurate matching of demand, but also makes the distribution strategies more fair in peak hours.

2.2.2 Anticipating and foreseeing Systems Dynamics

Renewable era forecasting is not much about precision but is more of adaptive anticipation. Since renewable energy can vary with the natural cycles, the systems need to constantly learn and adapt to the unpredictability. Predictive maintenance models have in turn embraced the concept of probabilistic reasoning that predicts the probability of component failure relative to stressors in the environment as opposed to absolute time limits.

Systemic forecasting is beyond maintenance and connects weather intelligence with behavioural response. As an example, now grid managers can pre-empt energy shortages by inducing dynamic tariffs or voluntary demand response events prior to the occurrence of imbalances. These predictive models are based on multiscale data of local microclimates, regional generation prognostications, and even indicators of the economy to coordinate stability in an increasingly decentralized network (IRENA, 2022).

2.2.3 Optimization and Control: The New Logic of Balance.

Renewable energy system optimization should now be used to address various purposes: energy security, cost-effectiveness, and environmental sustainability. Multi objective optimization, which is a trade-off between short term grid reliability and long-term carbon reduction, is being designed to be implemented by intelligent control systems.

Dynamic storage dispatch algorithms decide when and where the energy should be stored or released to minimize curtailment and maintain stability of the grid frequencies. The development of peer to peer (P2P) trading platforms enables community groups to trade in excess power, and it makes consumers the market participants. This empowers resiliency with social equity being instilled in technical operations. On the systemic level, optimization turns out to be an ethical practice that makes computational efficiency to be in line with environmental justice in addition to automation complementing decision making by humans as opposed to displacing them.

2.2.4 Feedback and Regeneration: Learning through the Energy Flows.

Possibly the most radical aspect of this new paradigm is the feedback learning systems which improve over time through a self-examination of their performance in terms of ecological and social measures. Instead of just measuring voltage stability or downtimes, grid operators are also measuring efficiency of emission reduction, indices of energy equity and environmental regeneration scores. These metrics are a loop to governance and algorithm optimism, to make sure that technology interventions are not growing beyond the planetary limits. In this regard, the energy system itself can be regarded as a learning ecosystem, which can optimize its structure and functioning in search of shared resilience as opposed to the efficiency of what is narrowly defined.

2.3 Challenges

Although renewable intelligence is indeed the future, there are important issues to consider making the low carbon transition sustainable and fair.

2.3.1 Computational Energy Footprint.

Ironically, the systems that are aimed at being more efficient in energy managing can become large consumers of energy themselves. According to Strubell et al. (2020), the growing proportion of the world electricity is already attributed to intensive computing workloads linked to training and executing analytical models. The digital infrastructure of the energy transition may undermine the gains it is supposed to bring, without an adequate design. New research proposes energy conscious computation, application of model compression, distributed inference, and data centres that are powered by renewable energy to alleviate this paradox.

2.3.2 Integration Complexity

The second major challenge is coordination at multiple levels: technological, behavioural and regulatory. Renewable facilities need real-time coordination among physical infrastructure, human intervention and market mechanism. For instance, the generation patterns have to be compatible with dynamic prices and user flexibility. This demands a different form of socio technical literacy in which, energy consumers are not passive end users of the system, but they are also co-creators of system stability.

2.3.3 Governance and Equity

Technology cannot be a successful step to the renewable transition. Even the most efficient systems can make the problem of inequality or environmental deterioration persist through the failure of governance systems to guarantee equitable access, data transparency, and accountability. Building sustainable intelligence in the energy sector should therefore be designed to have an ethical supervision and inclusive governance that will balance innovation and justice.

3. RESILIENT CITIES AND MOBILITY SYSTEMS: DESIGNING THE URBAN EARTH

The Anthropocene thick ecosystems in which economic action, sociality, and environmental influence are nexus are characterized by urban areas. By 2050, close to 70% of the world population will have moved to cities (Nkengla-Asi et al., 2024) however, most urban systems will be fragile, with linear flow of resources, overloaded transport systems, and infrastructures at risk of severe weather. The future of the planetary resilience, thus, will be determined in the architecture of cities: their movement, respiration, and adaptation.

Resilient cities are not necessarily technological smart cities. They are dynamic systems that can learn through stress and redistribute with a regeneration of ecological health. In the new urban paradigm, we no longer see infrastructure as a static concept, but rather as fluid organs that react to feedback from environment and human beings, such as streets, grids and buildings. Data and sensing technologies are not surveillance tools, but a tool of stewardship, making all people aware of the air quality, movement, and resource circulation.

The section examines the ability of integrated intelligence to develop urban resilience by redesigning mobility, symbiosis of infrastructures and climate responsive planning and in the context of the ethical, social and ecological dimensions defining the urban future

3.1 Urban Complexity: The Evolving Logic of Urban Systems

Traditional cities were based on predictability (fixed routes, centralized utilities, and linear growth). But modern urban existence is in motion. Migration, climate change and changing consumption patterns pose lively stressors which are not handled by the rigid systems.

The resilience of urban design starts with nonlinearity: floods and droughts have irregular patterns; energy and transport needs vary with the weather and social behaviour; and the social inequalities influence access to mobility and green lands. The following generation of urban systems will then need not only sensors and analytics, but contextual intelligence the ability to perceive patterns, change based on them, and continuously develop.

Mobility, water, waste, energy and housing represent the main systems that are interconnected to create an urban system that is resilient. These systems enhance vulnerability when designed independently and increase resilience when connected to each other by adaptive coordination. As an example, flood control network of a city may relate to mobility routing framework, which will redirect transport in real time and restore groundwater using green corridors. The logic of the 21st century city, therefore, changes from that of optimization for efficiency to orchestration for balance.

3.2 Mobility Redone: The Mobility turned into Regeneration

The urban metabolism is centered around mobility. For decades transportation planning has been concerned with throughput - how to move the most people or goods in a fixed amount of space. However, such efficiency-based solution usually creates contradictions: traffic jams, pollution, and unfair distribution. The subsequent step of urban mobility reinvents movement as a rejuvenating transformation that limits emissions but recovers the urban health and social unity.

3.2.1 Decarbonized Transport Ecosystems

The electrification of the mobility has gone much further than electric vehicles It consists of the modal integration, which is a networking of walking, cycling, autonomous shuttles, and shared electric fleets. Smart charging roads and transport hubs that are integrated with a grid enable vehicles to become mobile sources of energy, which can store and release electricity back to the grid when there is a peak of demand. Adaptive design could even be used in terms of transport and water management in a single area in coastal cities or flood prone cities to combine floating transit corridors or amphibious paths to function as flood buffers.

3.2.2 Data Informed Equilibrium

Urban traffic is slowly no longer a fixed flow challenge but a real-time negotiation among a huge number of agents' vehicles, pedestrians, micro mobility devices, and so forth. Information integration centres are now combining real-time data of road sensors, transit, and atmospheric monitoring stations towards flow optimization and minimization of exposure to pollution. But not only the algorithms, but the policy intelligence of governance frameworks to change incentives and design decisions in response to changing conditions is what determines resilient mobility. Adaptive congestion pricing, flexible zoning and the mandate of equitable access will all ensure that the efficiency gains don’t come at the expense of inclusion and environmental integrity.

3.2.3 Human Centred Mobility

Finally, the resilient mobility systems need to relate to the human rhythms to create the cities where walking, community, and slow mobility can co-exist with high-speed connectivity. A new model of reliance on urban resilience is the design of the 15-minute cities in which the main services are within the walking or cycling range. When movement becomes a source of well-being, and no longer a source of stress, cities become not a network, but a system of circulation of life.

3.3 Climate Responsive Urban Design

The next generation of cities will have to act as climate moderators and not climate victims. Conventional urban planning of concrete, asphalt and vertical growth has increased the urban heat island effect and disrupted the water cycles in the local area. Climate responsive design on the other hand combines green blue infrastructure, circular material flows and renewable powered districts to align the built and natural environment.

3.3.1 **Ecological Integration**

The surfaces of the city are being redefined as part of the processes in the environment. Green roofs, porous pavements, and bio swales minimize runoffs and restore ecological diversity. Bio digital twins are now being incorporated into some cities to provide simulated digital representations of natural systems that model the response of vegetation or soil changes on microclimates and carbon sequestration.

3.3.2 Urban Energy Symbiosis

Buildings, traditionally the largest contributors to urban emissions, can transform themselves into energy symbiotic nodes collecting solar power, storing energy and sharing it with surrounding buildings. New prosumer districts already exhibit group energy resiliency using cooperative microgrids that match production and consumption on-the-fly.

3.3.3 **Social Dimensions of Resilience**

Physical resiliency cannot be separated from social resiliency. Access to green infrastructure, cooling centres and clean transport need to transcend socio economic boundaries. Without inclusion, urban adaptation risks exacerbating inequities, exposing vulnerable communities to the adverse effects of climate hazards. Thus, resilience planning requires a justice centred lens where planning for design is about supporting not only survival, but about dignity and opportunity as well.

4. AGRICULTURE, WATER, AND ECOSYSTEM MONITORING: CULTIVATING INTELLIGENCE IN LIVING SYSTEMS

Agriculture as well as ecosystems define the biospheric foundation on which all economies and civilizations depend. Yet they are among the most vulnerable to increasing climate uncertainty. Rising temperatures, changing rainfall patterns and soil degradation endanger the world's crops and will cut production by up to 20% by mid-century (Abebaw, 2025). At the same time, freshwater scarcity and biodiversity loss are hitting critical tipping points that are undermining food security and ecological resilience.

The challenge here is not just to make more with less, but to match productivity with planetary health growing systems that regenerate rather than deplete. New data ecosystems and analysis frameworks can monitor, predict and synchronise agricultural and hydrological processes at unprecedented scales. But deeper than the digital layer are all the other transformations that depend on a far greater re-orientation: that of seeing farmlands, watersheds, and forests not as individual units of production, but as living, learning networks within the metabolism of the Earth. In this sense, sustainable intelligence in the field of agriculture and water systems must be able to connect three dimensions: precision, participation, and planetary coherence. It must sense ecological rhythms, change in climates, respect the social knowledge in local communities

4.1 Sectoral Pressures: The Converging Crises of Food, Water, and Ecology

Modern agriculture works in an intersection between ecological stress and economic dependency. Global food systems use 70% of freshwater withdrawals (Solaw, 2011) account for almost one fourth of greenhouse gas emissions (Poore & Nemecek, 2018) and extensive habitat conversion. The increasing demand for protein and volatile climate patterns are causing enormous pressure on soil fertility and water balance.

Conventional intensification through monoculture and chemical dependence has brought short-term productivity but long-term vulnerability; It is soils and aquifers that provide the resilience of the systems to which they belong, but they are now becoming degraded. As a result, a new paradigm is developing - regenerative agriculture, which is based on soil health, carbon sequestration and biodiversity restoration while not compromising productivity.

Within this paradigm, adaptive technologies are not so much substitutes for ecological intelligence as exponents of natural feedback. Soil moisture monitoring, plant physiological indicators of stress and hydrological runoff prediction could match agricultural management to environmental variability in such a way that a resilient trade-off of yield and renewal would result.

4.2 Applications: Sensing, Modelling, and Regeneration

The intersection of Earth observation with environmental modelling and participatory knowledge networks helps to achieve the multi scalar transformation from an individual farm plot to the whole bioregion

4.2.1 Precision Farming: From Control to Care

Precision agriculture has branched out of efficiency to include stewardship. Networks of sensors in the soil and on crops, combined with high resolution satellite imagery can monitor the levels of nutrients in plants, the rate of evapotranspiration and pest pressures in near real time. However, the real power lies in being able to translate these observations into context aware interventions that can cooperate with ecological cycles instead of overriding them. Machine learning models can be used to map variability in soil carbon, to forecast pest outbreaks and some of the suggested interventions are low impact interventions such as biological pest control or regenerative cropping sequences. These models have captured over 55% of variance in sustainable farm outcomes, with local experience calibration, confirming just how effective coupling data precision and field knowledge could be in farming practice.

4.2.2 Water Systems: Dynamic Equilibrium Management

Water systems are biosphere circulations. The key to future sustainability is to treat them as commons that can be managed flexibly, rather than utilities that cannot be seen as commons. Predictive hydrological models can anticipate the onset of drought, detect leaks or inefficiencies in irrigation or to guide coordination at the watershed scale between farmers, utilities and municipalities.

Integrated water balance monitoring systems use combinations of ground water sensors, remote sensing, and people-based monitoring systems to ensure ongoing and real-time water balance. These tools can help ensure an even distribution in times of scarcity and avoid overharvest by allowing a visualization of the environmental impact of the current consumption habits. By linking farmers decision making and the basin level water information, the allocation of resources becomes not only efficient but ethically transparent an important step towards long term hydrological justice.

4.2.3 Ecosystem Monitoring: The Living Earth Observatory

Conservation has become based upon distributed observation networks which serve as a kind of planetary nervous system. Camera traps, sound recorders, hyperspectral satellites reveal changes in vegetation, migration routes or species loss with ever higher resolutions. These systems can detect illegal deforestation, habitat encroachment or biodiversity decline before irreparable damage has been done. However, ecological agency (the capacity of local and global actors to act together) is needed to augment monitoring. Future monitoring systems need to link detection with governance - making sure that alerts will lead to action, not indifference. This needs transparency, free access to data, trust and collaboration among institutions and communities.

4.3 Trust, Ethics, and Local Knowledge

Technological advance is impossible in the absence of human wisdom and social legitimacy. Water and agricultural innovations must be integrated in ecosystems based on participation that considers local knowledge, customary rights and traditional ecological practices.

Multi objective decision frameworks (enforcement learning for resources optimisation) can balance productivity, equity and environmental stability. Yet these systems need to be explainable and transparent and enable farmers and local authorities to understand and challenge their logic and to adapt it.

No less important is the ethical side: Who owns the data, who can profit from predictive knowledge and how do cares for trust in areas where the digital divides are still on-going? Participatory governance of combining open-source infrastructure with local co design will build accountability and strengthen the legitimacy of sustainable transitions. When knowledge flows in both directions technological and traditional systems evolve from smart to wise. Wisdom, in this context, is defined as the integration of innovation and integrity of place, people and planet.

5. DISASTER RESPONSE AND CLIMATE RISK MANAGEMENT: ANTICIPATING THE UNPREDICTABLE

The twenty first century is a time period of continual uncertainty. Climate related disasters considered as rare a decade ago are now happening with more occurrence and intensity. The Climate Risk Index 2025 report records that between 1993 and 2022, extreme weather events caused inflation-adjusted losses of nearly $4.2 trillion caused by over 9,400 extreme weather events (Adil et al., 2025). Beyond infrastructure damage, these crises destroy livelihoods, widen the gap in inequality and push and test the resilience of ecosystems and governance alike.

The existing disaster management systems, based on static risk maps and reactive management, are not suitable for a fast-changing climate. What is emerging instead is a new paradigm of anticipatory resilience a distributed, data informed architecture that can detect weak signals, simulating cascading risks, and orchestrating real time response.

Within this framework, sustainable intelligence does not constitute a collection of algorithms but a living coordination system one which perceives the initial warning signs, and provokes social action on an individual scale or a large scale and then learns from each event to optimize its resilience in the future.

5.1 Predictive Analytics: From Forecasting to Foresight

Modern day disaster preparedness starts with early warning, but what really changed is early learning. Predictive analytics have grown from simple hazard predictions to multiple-dimensional prediction systems, combining information sources such as climate projections, social vulnerability indices and ecological thresholds.

Contemporary models take the input of satellite imagery, hydrological simulations and sensor information and identify anomalies well before people can see them. For example, by examining radar-based measurements of soil moisture activity, this has the potential to detect precursors to landslides; by using spectral imaging to spot abiotic stress (e.g. vegetation) that precedes a wildfire; and by using models of the ocean to track the development of cyclones weeks in advance.

More importantly, predictive analytics are becoming less descriptive ("what happened") and more prescriptive ("what should we do now") systems. Instead of issuing alerts alone, they can now simulate different evacuation routes, the availability of resources and recommend adrenal responses. "This means that in each disaster, a collective memory has grown that helps make the world more prepared, and that the models learn from each disaster too."

5.2 Integration with Other Systems: Building Adaptive Infrastructures

Disasters seldom respect geographical, institutional or technological boundaries. Hence, resilience cannot be developed out of isolated silos. Rather, it must result from systemic integration between urban, energy, agricultural and social infrastructures.

For instance, models of dynamic evacuation in urban areas interact with smart mobility, modifying the flow of traffic in real time, so that the movement is faster toward safety areas. In rural applications, the hydrological sensors combined with irrigation networks can find automatic valves to close them to prevent contamination in case of floods. Similarly, grid aware infrastructures can balance the power distribution, providing protection to critical infrastructures like hospitals and emergency communication during extreme heat events.

Cross sector integration also makes it possible to understand compound risk modelling: how one hazard exacerbates another. For example, heatwaves can lead to higher energy demand, which can be a strain on grids and lead to blackouts, further deleterious to people's health. Integrating these interactions within predictive systems assists policy makers in designing multi hazard resilience frameworks to turn fragmented responses into coordinated adaptation.

5.3 Human Dimensions and Ethical Governance

No prediction system can work without human trust. Effective disaster risk management demands free access to governance, local participation and ethical management of data. Early warning networks will need to focus on inclusivity that ensures the marginalized, who are often the most vulnerable populations receive information in accessible formats and languages. Thirdly, data driven decision making needs to be socially aware: automated alerts that lack local traditions or realities of evacuation can have the opposite effect of causing confusion or resistance. Community based sensing and feedback systems could close this gap combining digital accuracy with human situations.

Ethical frameworks for governance based on principles of privacy, fairness, and accountability should govern the collection, sharing, and action of information. This is particularly important in crisis contexts, where the misuse or misinterpretation of data can have the effect of making an already terrible trauma even worse. At the end of the day disaster response is a test of collective intelligence: the capacity of human and technical systems to work together under pressure and spend their time adapting to new realties and preparing for resilience into the future.

5.4 Approaching Regenerative Resilience

Beyond that of preventing loss, the next generation of disaster intelligence aspires to a purpose of regenerative recovery the ability to grow stronger from the disruption.

This principle is homologous to natural systems, where disturbance is more often the cause of adaptation and renewal. Rebuilding efforts can therefore offer an opportunity to decarbonised energy systems, restore ecosystems and strengthen social equity. Environmentally informed post-disaster reconstruction can lead to the use of low carbon materials, nature-based flood defences or community led green buffers, which can serve as biodiversity corridors. In this understanding of resilience, it is a constant feedback loop - anticipate, act, recover and learn. The focus then becomes not so much a matter of survival, but adaptive flourishing, where everything happened yields the knowledge for the long-term adaptation betoken of the stability of the planet.

6. INDUSTRY, CIRCULARITY, AND RESOURCE REGENERATION: DESIGNING FOR PERPETUAL VALUE

Industrial systems, which are responsible for almost a third of world-wide greenhouse gas emissions, are now entering a phase of structural redesign. The traditional logic of economic activity bone-deep linear logic of take-make-dispose-has reached ecological and economic limits. Global pressures on critical minerals, the growth of waste amounts and increased production costs are a sign of the fragility of resource-intensive growth.

The future of production is in closed loop self-correcting ecosystems where materials, energy, and information circulate as part of a network supported by the sectors. In this configuration, data driven intelligence is not just a tool used to make things smart dermal optimization but rather a connective tissue that is designed to connect materials, products and services through feedback and regeneration. Circularity thus develops beyond recycling but becomes a systemic discipline, inwards to design and manufacturing, to create a commuting design cycle together with manufacturing, logistics and consumer behaviour-No Planet B The ultimate aim is to ensure that materials are used as much as possible and waste is reduced as much as possible, not at the end of the process but at the start.

6.1 The Circular Economy: Beyond Efficiency

The circular economy is a transformative industrial paradigm one that repositions progress in terms of regenerative throughput instead of extractive output. As the Ellen MacArthur Foundation (2020) points out, decoupling economic growth from resource consumption and pollution is achieved in circular systems through the design-out of waste; designing products to keep them in use; and regenerating natural systems.

Circular models are becoming part of manufacturing, textiles, construction and electronics. For instance, materials passports are used to track the composition, origin and lifecycle of industrial components to allow manufacturers to recover or re-use assets at end of life. Product as a service models have an opportunity to move fulfilment from consumer to producer, and the resulting need for a longer working life and repair will be incentivised.

Bio based and recyclable materials are substituted for resource intensive composites and distributed ledger systems are used that keep all supply chains transparent. Crucially though, circularity is growing from niche importance as a sustainability strategy to a macroeconomic framework where resource stewardship is a part of national industrial policy. Countries such as Finland, Japan and Netherlands countries have developed “Circular Economy Roadmaps” that, among industrial design, waste policy and funding of innovation, are all part of coordinated systems.

6.2 Intelligent Applications and Industrial Synergies

6.2.1 Digital Twins and Systems Simulation

Digital Twins and Digital Twins of Systems and Systems Simulation Digitally twins of industrial ecosystems enabled by virtual objects allow for constant analysis of environmental and material flows. These systems reflect real world operations and allow engineers to simulate the impact of design changes, forecast the need for maintenance and simulate material flows before real interventions take place. By combining operational data with environmental parameters, digital twins help to achieve an optimized use of energy, suitability of materials and thus the real-time recognition of bottlenecks.

6.2.2. Predictive Maintenance and Material Longevity

Predictive maintenance is a type of maintenance system that can detect worsening degradation patterns and material fatigue so that timely repairs can be carried out and equipment components reuse can be made possible. Manufacturing interventions include datapoint monitoring by certain equipment and lengthening the lifecycle of equipment by up to 30 percent; eliminating both downtime and waste. When extended to industrial networks, diagnostic repair delivers a real positive contribution to conserving resources and being carbon neutral.

6.2.3 Supply Chain Transparency and Regenerative Sourcing

One opportunity for regenerative transformation which is perhaps its greatest untapped asset is the hidden backbone of industrial activity, the supply chain. New transparency frameworks combine material provenance, carbon intensity and social responsibility indicators in unified dashboards. These enable organizations to make decisions about suppliers not just on the basis of cost, but also on the suitability in terms of circular compatibility. Such a level of traceability provides the assurance that recycled, or ethically sourced materials will be able to be seamlessly recycled through production cycles with a new level of accountability to the world as a whole.

6.3 From Optimization to Regeneration

While efficiency makes the least of waste, regeneration gives back value. And the next industrial frontier goes beyond the conceptualization of minimization of impact and fundamentally deals with ecological and social system healing. For instance, industrial symbiosis clusters co-locate manufacturing plants to turn the waste of one company into an input for another (using emissions as feedstock for another company or residual heat to produce district energy, for example).

Material innovation is also changing the industrial metabolism: biodegradable polymer, low carbon concrete, and carbon negative composites enable factories could successfully turn into carbon sinks instead of being sources of carbon. Paired with energy systems that are powered by renewables and localised microgrids, these developments reimagine production as something that is part of earth's regenerative cycle, rather than an extractive chain. The fact that there is the integration of circular metrics into corporate governance further leads to an acceleration of this transition. Lifecycle accounting, dynamic carbon pricing and embedded sustainability auditing; Making regeneration a measurable business outcome, instead of a wish.

A schematic illustrating the flow of material and information cycles within an industrial ecosystem in Figure 3. The continuous feedback loop ensures perpetual resource circulation and regenerative industrial design.

Figure 3. Circularity Intelligence Architecture

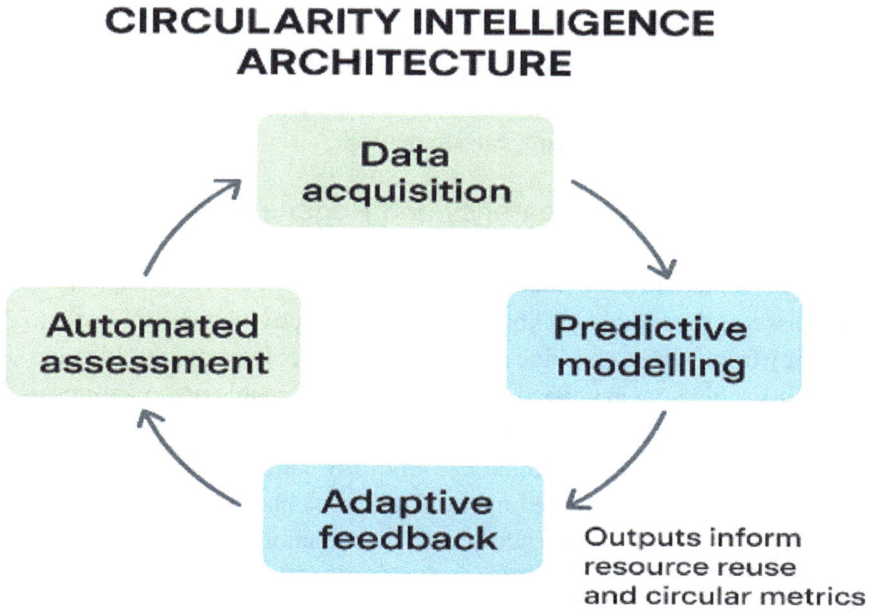

7. CROSS DOMAIN INTEGRATION AND GOVERNANCE: DESIGNING THE PLANETARY OPERATING FRAMEWORK

The deep transformation of civilization on a sustainable basis will not be possible on a piecemeal basis. Energy, agriculture, water, infrastructure and industry are massively interconnected systems, interacting with one another through materials, data and influence both in space and time. The challenge for the coming decades is that structures in governance and integration need to be established and created in such a way that these interdependencies are recognised and handled in real-time. While digitized infrastructures now produce enormous quantities of data from all sectors, most still exist in silos: energy networks, urban transportation, agricultural databases and industrial supply chains. To pave the way for a planetary intelligence, adaptive and shared architecture that allows the free flow of information between domains of action, is necessary, that is one of the conditions for the development of foresight and collective action.

7.1 Cross Domain Integration: Building the Planetary Fabric

The future of sustainability lies in interoperable intelligence systems designed not as isolated tools but as communicating ecologies. In this vision, distributed networks of sensors, models, and governance mechanisms form a single "planetary fabric" that perceives and responds to environmental, social, and economic change holistically.

7.1.1 Unified Data Lakes and Dynamic Modelling

United data lakes bring together data from all industries: meteorological observations, industrial euro thermometer flows, agricultural yield and city traffic, etc. into a common analytic middleware. These repositories make it possible to do cross sector modelling, and thus to identify cascading dependencies not otherwise visible. For example, a water decrease in rivers, which is detected by hydrological sensors could automatically start adjustments in hydropower generation, irrigation schedules and regional transport logistics.

7.1.2 Shared Decision Engines and Adaptive Coordination

Lying on top of such data infrastructures, multi-domain orchestration platforms with shared decision engines are able to balance conflicting goals like energy reliability, food security and ecological preservation. Rather than optimizing any single sector, these engines take planetary objectives into account: optimize carbon emissions, preserve biodiversity, and maintain social equity. This idea is based upon the notions of natural ecosystems and is one of the facets of feedback which affects the stability of the system. As digital twins of whole economies become a possibility, such platforms have the power to simulate the results of policies before they are implemented, lessening the risk that policies will fail, and ensuring evidence based, anticipatory, governance.

7.1.3 Planetary Metrics and Interoperability

Cross domain integration requires a value measure commonality. The future sustainability frameworks are expected to be based on planetary metrics dynamic indicators that translate the information on carbon, water, biodiversity, and human wellbeing into integrated dashboards. These indicators are conceived as feedback loops for governments and organisations not only to ensure that action is informed and directed in real time but to ensure that government and other organisations are consistent with global thresholds, such as the Paris Agreement and the Sustainable Development Goals (SDGs).

7.2 Governance: Ethics, Oversight, and Human Agency

The combination of technology and governance suggests the potential of increasing inequalities and the acceleration of environmental deterioration. Sustainable intelligence therefore needs to develop within a governance architecture which is based on ethics, transparency and participation. The point is not to control (versus complex systems), but to govern (versus life), and to use intelligence in a manner that ensures the perpetuation of life and not the perpetuation of power.

7.2.1 Ethical Foundations

Values are the foundation for governance. Systems that would influence planetary processes have to reflect notions of equity, inclusiveness, and environmentally responsible people. Combining concepts in environmental justice, human rights, and indigenous knowledge, values and norms should inform design principles that create decision engines, model outputs, and data collection that east not violate human and non-human stakeholders.

7.2.2 Regulatory Alignment and Reflexive Law.

Static regulation cannot even keep up with emerging technologies. Emerging theories of adaptive governance suggest a feedback-based regulation in which policy becomes self-aware and adaptive (agenda 2024) along with technological systems. Such frameworks can integrate environmental law, data policy and industrial standards into logical mechanisms that develop in an organic way. Cross border agreements could help institutionalise planetary scale carbon, material and data ethics accountability.

7.2.3 Human–Machine Collaboration

True sustainable intelligence must preserve human intentionality in automated systems. Individuals and groups articulate their societal goals, ethical limits and collective aspirations; intelligent systems solve them into evidence-based adaptive action. This relationship reflects the so-called "human policy loop" in which human deliberation and technological feedback loops back and forth. Rather than replacing the judgment of human decision makers, automation should be an augmentation of planetary stewardship, empowering communities and policymakers with forethought, precision and scenario prospect. It follows that the key to governance is not substitution, but symbiosis.

7.2.4 Governance Networks for Planet Earth

In the long term, governance may manifest itself as a federation of planetary councils distributed networks working to co-ordinate environmental, technological and social data across the regions. These institutions would act as digital ecosystems of accountability, which would entail making sure sustainability thresholds are met - not only at the national level but at a systemic one as well.

8. CHALLENGES, ETHICS, AND EQUITY: DESIGNING INTELLIGENCE FOR A JUST PLANET

The accumulation of smart systems into planetary infrastructure is resulting in not only opportunities for sustainability as well as the delivery of deep ethical, social and ecological dilemmas. As these systems start to be applied in critical areas (energy distribution, access to resources, agricultural productivity, and environmental regeneration) questions such as whether intelligence can optimise, become constrained by the most important questions of who benefits and who is excluded, and what are the driver values of the optimisation? As a concept, sustainable intelligence thus necessarily needs to be grounded on an ethical backbone as strong as its technical one. Without this grounding, the very tools used to mitigate the risks of climate change, and social risks, are liable to increase those risks.

8.1 The Energy Footprint Paradox

The irony of the modern computational development is the superhuman requirement of energy by intelligence itself. Power consumption of global data centres will more than double from 683 TWh in 2024 to 1,479 TWH by 2030 as pointed out by ABI Research's latest analysis, which provides the most extensive global report on data centres. Each process of training, each run of optimization is part of an unperceived carbon shadow: a paradox that, though printing is done to fight depletion, it reveals that resource-intensive processes are needed.

Green computing strategies of computation provide a partial solution. Large-scale models' environmental footprint can be substantially reduced through the implementation of model compression, low energy architectures, renewable powering of data centre operations, and other "energy aware" algorithms. Yet true mitigation requires systemic awareness of a redesign of the digital economy to ensure that computation is scaled in harmony with planetary boundaries.

Emerging research has a vision of distributed, circular models of computation, in which localised nodes, powered by renewable microgrids, replace energy intensive central servers. Such an approach might democratize free access, increase resilience and bring sustainable thinking to the infrastructure of digital intelligence itself.

8.2 Equity and Inclusion: Avoiding Algorithmic Asymmetry

The next large frontier is equity. Intelligent systems can have the effect of worsening global inequalities if these systems advantage the data-rich and technologically advanced northern nations while disadvantaging the Global South, rural areas or underrepresented groups. In many environmental blocs, this lack of data means that the experience becomes an injustice within the scientific community, as underserved communities on geographic and economic terms face fewer predictive warnings, less prioritized resources and have slower recovery times after climate-related events.

In order to address this imbalance, there needs to be a change from the rhetoric of "AI for all" to the practice of knowledge equity. Objectives: Community-based and participatory design, open data creation, and public participation guarantee that people most affected by ecological degradation have a significant role in designing the intended technologies to benefit them. In practice very similar to localised sensing networks sustained by citizen scientists, region specific models produced in conjunction with local institutions, and global entities under the rule of multi stakeholder coalitions instead of single corporations. The objective is not just technological inclusion but epistemic justice, keeping in view the fact that sustainability knowledge must emerge out of different cultural, geographical and ecological conditions.

8.3 Ethical Dilemmas and Value Frameworks

Sustainability choices are moral choices: if we have a finite number of resources, every optimization produces an exchange. Algorithms that operate energy grids, agricultural cycles, or urban infrastructure, therefore, must deal with ethical tensions under uncertainty decisions that have implications for livelihoods, ecosystems and even future generations. For instance, a focus on meeting the immediate energy needs of the human species above ecosystem restoration could have short term benefits but long-term ecological destruction as a result. On the other hand, existing conservation policies can limit rate of development in vulnerable regions. These are not computational problems per se, but philosophic negotiations that are embedded in system objectives.

Other ethical models like the "principles of planetary justice" and multi criteria decision ethics suggest practical ways forward. They propose that the decisions at system-level should incorporate explicit moral constraints such as intergenerational equity, ecological preservation and distributive fairness. Opaque auditing and explainable decision-making means that not only does the trade-off appear but it is visible, discussed, and democratically authorized. To deal with these predicaments, future governance designs will need ethical feedback loops: efficient systems by which human supervision repeatedly assesses the implications of automated decisions.

8.4 From Governance to Stewardship

Ultimately, the challenge of ethics in planetary systems transcends compliance and enters the realm of stewardship a deep moral responsibility to act as caretakers rather than controllers. The notion of stewardship reframes humanity's relationship with technology and the environment: intelligence should not dominate natural systems but participate in their regeneration.

This shift implies a cultural evolution in how we define progress. Rather than measuring success by speed, scale, or economic return, future systems must evaluate themselves through planetary indicators resilience, biodiversity, equity, and wellbeing.

Ethics, therefore, is not an afterthought or an overlay; it is the design principle that determines whether intelligent systems contribute to planetary stability or accelerate its decline. Sustainable intelligence demands nothing less than moral imagination at a planetary scale.

9.0 THE ROAD AHEAD: SCALING REGENERATIVE INTELLIGENCE

The transition from isolated innovation to planetary scale regeneration marks the next defining leap for human civilization. The future no longer belongs to intelligent tools operating in silos but to integrated ecosystems of intelligence that align economic growth, environmental resilience, and human flourishing within shared planetary limits.

As the global community faces converging crises climate volatility, biodiversity loss, and social fragmentation intelligence must evolve from optimization to restoration, from control to coexistence, and from extraction to reciprocity. The "road ahead" thus demands not only better systems, but wiser ones capable of learning from the planet rather than merely computing upon it.

9.1 Emerging Trends: Toward a Regenerative Paradigm

Recent years have witnessed the rise of Green and Regenerative Intelligence a paradigm shift that integrates sustainability directly into the design of digital systems. Green computation focuses on efficiency, while regenerative computation goes further, embedding principles of renewal and replenishment into algorithms and infrastructure.

Open standards and interoperable architectures are becoming essential for transparency and trust. As open-source ecosystems mature, the emphasis shifts from competitive ownership to collective stewardship a model that invites collaboration across governments, industries, and communities.

Moreover, multi-disciplinary deployment where ecological science meets data modelling, and ethics meets engineering represents the next frontier. The most resilient systems of the future will not be those that compute fastest, but those that think holistically, integrating planetary metabolism with digital infrastructure.

9.2 Systemic Scaling: From Pilots to Planetary Frameworks

Most sustainable technologies today exist as fragmented pilot projects localized, successful, yet disconnected. The true challenge lies in systemic scaling: building interoperable frameworks that allow insights, resources, and governance models to flow seamlessly across sectors and geographies.

Scaling regenerative intelligence requires a systems lens viewing climate response, agriculture, industry, and governance as interdependent organs within a single planetary organism. Unified data ecosystems and dynamic feedback loops will enable real time learning across domains, ensuring that lessons from one sector amplify benefits in others.

This transformation will also depend on policy coherence aligning environmental law, digital governance, and global trade with sustainability metrics. Public–private partnerships will play a pivotal role, but their effectiveness will rest on ethical grounding and transparent evaluation.

9.3 Collaborative Stewardship: Human–System Partnerships

The next generation of regenerative systems will thrive through collective authorship. Data scientists, ecologists, policymakers, local communities, and indigenous knowledge holders must become co designers of planetary solutions. This pluralistic collaboration ensures that technological progress remains grounded in ecological wisdom and cultural context.

Participatory governance frameworks are emerging globally to manage this convergence. They emphasize shared data ownership, community validation of models, and decentralized feedback mechanisms that allow regions to adapt technology to local needs.

Beyond coordination, the essence of collaboration lies in trust between nations, between sectors, and between humans and the intelligent systems they create. Such trust is sustained through transparency, accountability, and open dialogue about trade-offs and values.

9.4 Vision: From Smart Systems to Wise Civilizations

The ultimate horizon for regenerative intelligence transcends technology it envisions a civilization that learns as ecosystems do through balance, feedback, and adaptation. Intelligence, in this form, becomes a partner for planetary resilience, not a replacement for human judgment.

In such a world, the metric of progress shifts from GDP and throughput to ecological stability, community wellbeing, and intergenerational equity. Data flows resemble nutrient cycles reused, redistributed, and reinvested in the vitality of both people and planet.

This vision demands humility as much as innovation. The systems we build must mirror the principles of life itself: decentralization, diversity, and renewal. The future of intelligence is therefore not a machine that thinks like us but a system that thinks with us, sustaining the biosphere that sustains all.

10.0 CONCLUSION: FROM SMART SYSTEMS TO REGENERATIVE CIVILIZATIONS

Sustainable intelligence represents a fundamental transition from systems that are merely smart to those that are wise. The distinction lies not in computational power but in purpose in how intelligence aligns with the living systems it inhabits. While traditional approaches to innovation prioritized efficiency, productivity, and prediction, the next era demands a reorientation toward balance, reciprocity, and resilience.

The climate and ecological crises have revealed that fragmented progress no matter how technologically advanced cannot ensure survival if it neglects planetary boundaries. Sustainable intelligence, therefore, reframes technological evolution as an ecological relationship rather than a linear race for optimization. It calls for systems that not only adapt to environmental feedback but actively restore equilibrium, strengthening the resilience of both natural and social ecosystems.

10.1 The Evolution of Intelligence Systems

The concept of intelligence has historically evolved alongside human civilization from mechanical automation to digital cognition, and now to planetary scale awareness. In this new epoch, intelligence systems function as participants in the biosphere's feedback loops rather than as external observers. They learn from ecological rhythms, resource flows, and social signals, transforming information into restorative action.

This evolution signifies a departure from the extractive logic of industrial modernity. Instead of amplifying consumption, intelligent systems can nurture regeneration supporting biodiversity corridors, carbon negative industries, and equitable digital infrastructures. This paradigm envisions technology as a biocultural ally, integrating indigenous knowledge, local adaptation, and scientific reasoning into a coherent architecture of sustainability.

10.2 Governance and the Ethics of Continuity

No technological system, however advanced, can sustain planetary well-being without governance that embodies justice, transparency, and foresight. As intelligence systems grow more autonomous, the ethical scaffolding surrounding them becomes vital. Governance must evolve from reactive regulation to proactive stewardship, anticipating unintended consequences before they manifest.

This requires institutional frameworks that bridge data ethics, environmental law, and social inclusion. It also demands cultural change one that sees technology not as a replacement for human judgment but as an amplifier of collective wisdom. When integrated design meets ethical intention, sustainable intelligence becomes a continuum of care: between human innovation and Earth's regenerative capacity.

10.3 The Future of Regenerative Civilization

If the 20th century was defined by industrial acceleration, the 21st must be defined by planetary regeneration. The ultimate success of intelligence will not be measured by its computational reach but by its ecological coherence its ability to sustain complex life systems while uplifting human societies.

The future demands an ethic of technological humility: recognizing that intelligence, whether organic or digital, is part of a larger living system. The path ahead will depend on how societies choose to embed moral, ecological, and cultural values into the very code that drives their technologies. When done wisely, this integration will yield civilizations that learn from nature, act in reciprocity with it, and endure through its rhythms.

REFERENCES

Abebaw, S. E. (2025). A Global Review of the Impacts of Climate Change and Variability on Agricultural Productivity and Farmers' Adaptation Strategies. *Food Science & Nutrition, 13*(5), e70260. DOI: 10.1002/fsn3.70260 PMID: 40370417

Adil, L., Eckstein, D., Künzel, V., & Schäfer, L. (2025). Climate risk index 2025.

Agenda, R. G. (2024). Regulatory Experimentation: Moving ahead on the Agile Regulatory Governance Agenda.

Agyeman, J., Schlosberg, D., Craven, L., & Matthews, C. (2016). Trends and directions in environmental justice: From inequity to everyday life, community, and just sustainabilities. *Annual Review of Environment and Resources, 41*(1), 321–340. DOI: 10.1146/annurev-environ-110615-090052

Ali, D. M. T. E., Motuzienė, V., & Džiugaitė Tumėnienė, R. (2024). Ai driven innovations in building energy management systems: A review of potential applications and energy savings. *Energies, 17*(17), 4277. DOI: 10.3390/en17174277

Almond, R. E. A., Grooten, M., Juffe Bignoli, D., & Petersen, T. (2022). Living Planet Report 2022–Building a nature-positive society. *WWF: Gland, Switzerland.*

Fletcher, S., March, A. L. A., Roberts, K., Shirian, Y., i Canals, L. M., Cairns, A., ... & Bass, A. (2023). *Turning off the Tap: How the world can end plastic pollution and create a circular economy*. United Nations Environment Programme.

Floridi, L., & Cowls, J. (2022). A unified framework of five principles for AI in society. *Machine learning and the city: Applications in architecture and urban design*, 535-545.

International Renewable Energy Agency IRENA. (2022). *Renewable energy statistics 2022*. International Renewable Energy Agency IRENA.

Johnstone, I.Organisation for Economic Co-operation and Development (OECD). (2022). 10. Organisation for Economic Co-operation and Development (OECD). *Yearbook of International Environmental Law, 33*(1), 271–276. DOI: 10.1093/yiel/yvad054

Lal, R. (2020). Regenerative agriculture for food and climate. *Journal of Soil and Water Conservation, 75*(5), 123A–124A. DOI: 10.2489/jswc.2020.0620A

Macarthur, E. L. L. E. N., & Heading, H. E. A. D. I. N. G. (2019). How the circular economy tackles climate change. *Ellen MacArthur Found, 1*, 1 71.

Mukherji, A. (2023). Climate change 2023 synthesis report.

Nkengla-Asi, L., Bernardini, M. D. R. C., Cohen, M. J., Lawson-Lartego, L., & Coates, K. (2024). Sustainable Development Goal 11: Make cities inclusive, safe, resilient, and sustainable. In *Handbook on Public Policy and Food Security* (pp. 268-280). Edward Elgar Publishing.

Poore, J., & Nemecek, T. (2018). Reducing food's environmental impacts through producers and consumers. *Science*, *360*(6392), 987–992. DOI: 10.1126/science.aaq0216 PMID: 29853680

Roadmap, I. N. Z. (2023). *A Global Pathway to keep the 1.5° C Goal in Reach.* IEA.

Rolnick, D., Donti, P. L., Kaack, L. H., Kochanski, K., Lacoste, A., Sankaran, K., Ross, A. S., Milojevic-Dupont, N., Jaques, N., Waldman-Brown, A., Luccioni, A. S., Maharaj, T., Sherwin, E. D., Mukkavilli, S. K., Kording, K. P., Gomes, C. P., Ng, A. Y., Hassabis, D., Platt, J. C., & Bengio, Y. (2022). Tackling climate change with machine learning. *ACM Computing Surveys*, *55*(2), 1–96. DOI: 10.1145/3485128

Schwartz, R., Dodge, J., Smith, N. A., & Etzioni, O. (2020). Green ai. *Communications of the ACM*, *63*(12), 54–63. DOI: 10.1145/3381831

Solaw, F. A. O. (2011). *The state of the world's land and water resources for food and agriculture.*

Strubell, E., Ganesh, A., & McCallum, A. (2020, April). Energy and policy considerations for modern deep learning research. In *Proceedings of the AAAI conference on artificial intelligence* (Vol. 34, No. 09, pp. 13693 13696). DOI: 10.1609/aaai.v34i09.7123

United Nations Office for Disaster Risk Reduction. (2023). *Global Assessment Report on Disaster Risk Reduction 2023: Mapping Resilience for the Sustainable Development Goals.* Stylus Publishing, LLC.

Vinuesa, R., Azizpour, H., Leite, I., Balaam, M., Dignum, V., Domisch, S., Felländer, A., Langhans, S. D., Tegmark, M., & Fuso Nerini, F. (2020). The role of artificial intelligence in achieving the Sustainable Development Goals. *Nature Communications*, *11*(1), 233. DOI: 10.1038/s41467-019-14108-y PMID: 31932590

Wang, Z., Jang, W., Ruan, B., Wang, J., & Xiao, S. (2025). Developing and Integrating Trust Modeling into Multi Objective Reinforcement Learning for Intelligent Agricultural Management. *arXiv preprint arXiv:2505.10803.*

Chapter 7
Optimizing Weather Forecast Accuracy Through Eco-Friendly Machine Learning Techniques

Alaina Thea Concepcion
http://orcid.org/0009-0001-6874-6859
Bulacan State University, Philippines

Anthony Umali Concepcion
http://orcid.org/0009-0008-3634-2041
Bulacan State University, Philippines

ABSTRACT

This chapter discusses weather forecasting and Green AI by exploring the enhancement on how energy-efficient machine learning (ML) techniques can improve weather forecast accuracy while concurrently reducing carbon and energy footprints. The objectives are to: (1) evaluate and identify modern or advance energy-efficient ML models that make weather prediction provide accuracy with lower computational cost, (2) compare eco-friendly algorithmic approaches that is suitable for resource-constrained environments, (3) integrate sustainable data processing and storage practices into forecasting systems, and (4) assess and evaluate on how Green AI in enables on effective early warning systems for extreme weather events. By analyzing model-level strategies (pruning, quantization, distillation, and efficient architectures), and discussing systems-level interventions (edge inference, optimized data pipelines, and renewable-powered infrastructures).

DOI: 10.4018/979-8-3373-7554-0.ch007

INTRODUCTION

In latter years, weather prediction has become much more established than it was in the past in that it moved beyond a old-style statistical and physics-based approach of methodology, to advanced application of artificial intelligence (AI) and hybrid form of prediction or approach methodology. The first place forecasting systems were based on the application of dissimilar statistical models, such as regression analysis and time-series models that are not only very good in classifying past trends in weather conditions but are partial in the complexity and compound regulations of the atmosphere. The evolution and application of Numerical Weather Prediction (NWP) paved a pathway through which there is the use of mathematical models of atmospheric processes on a basis of the physical laws to forecast the future weather conditions. Although the models of NWP offer good systematically based projections, they become very dependent on initial conditions, computational resources as well as proper data adaptation methods to perform accurate forecasting. Previously, machine learning and deep learning model containing the application of artificial neural networks, and recurring architecture has also been developed as an actual tool that can learn nonlinear links with large and heterogeneous meteorological data sets. To present this section constraints of both sides of the argument as hybrid forecasting paradigms the incorporation of physics-based NWP performing with data-based AI models to improve predictive execution and resiliency have been instituted. The use of a set of predictions synthesizing the results of multiple models to measure uncertainty also offers additional support to decision-making in weather sensitive applications. These evolving paradigms offer a new step to more accurate, adaptive, and computationally efficient prediction systems that can be the core of the modern attempts to optimize the accuracy of the weather forecast.

Climate change has presented a lot of hesitation in weather patterns around the world, thereby ensuring a greater emphasis on the need to have more sophisticated and precise forecasting models. The antique numerical weather prediction (NWP) framework consumes energy as it implies large amounts of computational assets and energy that may surcharge environmental expenses by releasing carbon dioxide (Rolnick et al., 2022). The frames of Green Artificial Intelligence (AI) have developed in which the focus has been laid on building machine learning (ML) models that balance predictive performance and energy efficiency (Schwartz et al., 2020).

The focus of this chapter is to optimize weather forecast precision with the help of machine learning methods that are environmentally friendly at the expense of sustainable computing and environmental ethics. Additional insubstantial deep learning architecture integration and discussion, quantization, hybrid optimization and model pruning methods can significantly reduce the amount of computational energy required with no resulting accuracy loss (Xu et al., 2021). In addition to that can be played on data centers powered by renewable energy and green cloud computing glasses can be made to reduce the impact of AI-driven forecasting systems to the environment (Gupta and Singh, 2023).

This chapter aims to:

- Identify and analyze how energy-efficient machine learning models in the present picture enhance better weather prediction and reduce carbon impact.
- Examine and categorize eco-friendly view on algorithm which gathers prediction effectiveness but under resourceful computational facility.
- Integrate green processes of data managing and storing into prediction processes to promote the use of green AI.
- Determine how green AI can alleviate climate-associated risks in aiding the early warning processes of the extreme weather events.

With these in mind, the chapter subsidizes to the broad discussion on sustainable AI practices, points out that climate risk observing cannot be confined to making forecasts about environmental changes only, though it needs to minimize the digital carbon footprint of the predictive tools. Weather forecasting is very critical common good that supports and reinforces energy efficiency, agriculture, transport, and disaster preparedness. In previous years, brisk advancements in machine learning utilization have provided dissimilar models which rival on certain metrics, surpass conventional numerical weather forecasting (NWP) systems in version, as well as, in adjusting speed. However, these offer some benefit, which have accrued with the discussion compared to energy use and carbon emissions that come with training carried out an extensive application of machine learning models. The popularization of the new discipline of utilizing Green AI is intended to address the enhancement of the model in the area of the environmental impact through the advancement of energy-effective algorithms, minimal structure, and sustainable implementations. The approach and framework that brings the high forecast accuracy, with a reduced cost in energy and carbon in weather prediction tasks, is the subject of this chapter.

Machine learning-based weather prediction techniques vary between hybrid systems that unite both ML and physical systems as well as data-only deep learning systems that reproduce the behavior of the atmosphere. The major strategies of GraphCast, FourCastNet, Pangu-Weather, and GenCast, illustrate that ML can operate as dynamic forecasts at their orders-of-magnitude speedy period against the predictable NWP, and could reduce the energy path of the system forecasting procedures. These systems were under meticulous upgrades so as to recognize and accelerate the model computation within minutes or seconds on item equipment that accelerates decentralized fast forecasts to low resources environments. However, despite their promising that the application of deep-learning-oriented weather models can frequently demand but extensive available analytical sources during the coaching process, which might be based on cloud information structures that demand important lifecycle emissions. It is the theme of this chapter to convey this agitation.

This is dragging out on an already well-established and up-to-date body of literature to scale-up statistical weather forecasting, machine learning-based prediction, and Green Artificial Intelligence. Basic work is sum up and combined with the recent high impact work to supply some background on the field as well as the up-to-date interest. The chapter balances the developments in hybrid physics-AI models, grouped forecasting and energy-saving machine learning methods which in turn proves and talks of an effective conceptualise which helps in confirming the analytical and methodological addresses. Weather is the fundamental factor of the development of natural ecosystems, economic interactions, and human welfare. As the frequency of maximum weather outcomes heightened due to climate change, the necessity to enhance management related to weather has become as significant as at no point the before. Weather optimisation is not about holding weather as mentioned above, but rather a described process of the systematic utilisation of the improved technologies, forecasting models, and data-driven strategies to reduce risks and maximise the benefits linked to weather inconsistency. This trend is one which has transformed weather into an environmental factor that is reactive into a prohibition factor of strategy planning across an array of countries.

The reviewed literature presents current and primary studies in the field of meteorology. Through generating literature on numerical weather forecast, long neural networks, collective prediction, and Green AI values, the chapter forms a capable theoretical and practical base on how accurateness in predictions and computational sustainability can be mutually enhanced.

Background: Meteorological Forecasting and the Rise of Machine Learning

The chapter clearly illustrates the indicating issues, problems and the first currents modeling contemporary weather forecasting. Some of the main experiences that include the thrilling computational and energy demands of traditional numerical weather forecasting systems, vagueness dispersion, information scale, and carbon emissions linked to large-scale AI guidance are explicitly reviewed. In corresponding, the chapter attributes the most dominant trends entering the use and integration of machine learning that is founded on forecasting, hybrid physics of AI models, group systems, with edge and distributed suggestion, and the increasing implementation of Green AI ethics. All these factors serve to legitimize that prediction research is becoming the outcome of scientific accuracy and environmental responsibility. Part of the theme of the proposed book associated with placing weather forecast optimization in terms of its role in the grander scenario of a increasing climate variability, growing intensity of data, and increasing sustainability demand. It is associated to various central issues that are the restrictions of conventional numerical weather prediction models, the circulation of uncertainty, constraints of data quality, and the growing computational price connected with the higher resolution prediction. To counter these hindrances, the chapter presents the key tendencies that are conclusive in the profession, including artificial intelligence that gathers the techniques of predicting and combining them and Green AI principles to achieve a balance between accuracy and energy consumption. Such debates highlight the degree to which the paradigm of forecast is moving forward to deal with both science and pragmatic matters, and to support the significance of the chapter to the main theme of the book which is advanced and responsible technological solutions.

The chapter is firmly comprehensive that includes the methodological, computational, conservation, and operational scopes of the optimization of weather prediction. Instead of giving a tangible presentation of an imbalanced or peripheral issue, the discussion integrates the study of numerical modeling, machine learning novelty, the design of energy-saving algorithms, maintainable infrastructure, and real-world deployment concerns into one analytical framework. This integrative style also renders conceptual completeness at the same time as it is coherent to the major topic of the chapter that deals with the AI-enabled forecasting that is also green.

The Traditional Numerical Weather Prediction (NWP) and Its Costs

Numerical Weather Prediction (NWP) applies models that make high-hi-fi that prediction and forecasts constructed on first-principles physics by answering the atmospheric dynamics and thermodynamics equations on discretized grids. High-resolution NWP systems, showed by those run at nationwide meteorological cores, call for substantial high-performance computing (HPC) resources, rising in electricity intake for both simulation and data assimilation processes. The growth and operation of these structures also involve large data storage capacities and sustained power for observational networks. Due to the computational demands, many NWP centers operate large datacenters or supercomputing provisions that contribute to established energy budgets and greenhouse gas discharges.

Figure 1. Atmospheric model schematic showing climate interactions.

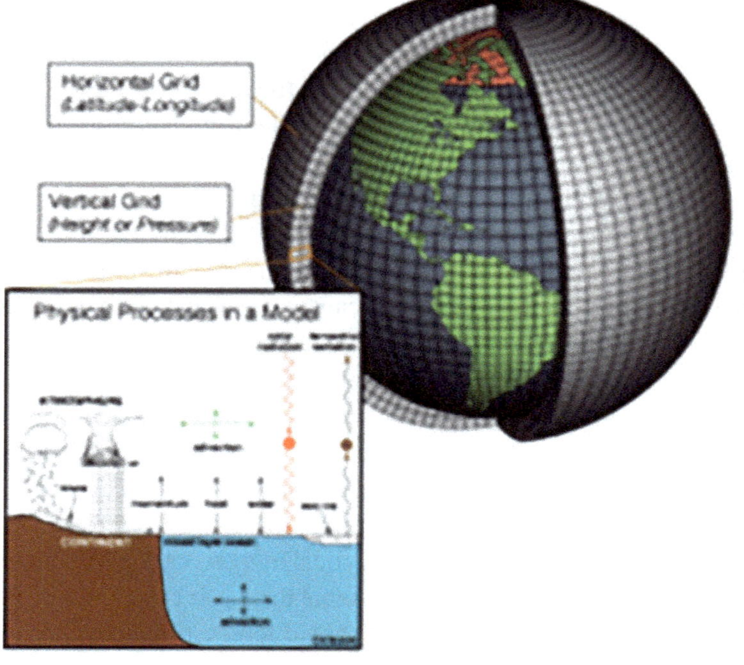

(Source: By NOAA, Public Domain, https://commons.wikimedia.org/w/index.php?curid=7263986)

Based on present atmospheric circumstances seen in Figure 1, Numerical Weather Prediction (NWP) uses specific mathematical models to prediction weather in terms of the atmosphere and oceans based on current atmospheric conditions (National Oceanic and Atmospheric Administration [NOAA], n.d.). Yet though NWP was first tried in the 1920s. Realistic predictions weren't made until the establishment of computer model in the 1950s (Lynch, 2008). At present, several global and regional prediction models manage global that apply data from radiosondes, weather satellite broadcasting, and other monitoring systems as key in (Encyclopedia, Britannica, 2024). The mathematical models are based on physical principles that use both short-term weather forecasting and longer-term weather for forecast on which this are essential for understanding and projecting climate change (NOAA, n.d.). Continuous advancements in regional models have significantly improved tropical cyclone track and air quality forecasts. But atmospheric models still face restrictions in precisely demonstrating localized events such as wildfires (Encyclopedia Britannica, 2024). Weather models are using a system that has a differential equations based on the study of laws of physics, using different tools like thermodynamics in predicting weather, radiative transfer to understand electromagnetic radiation, and atmospheric chemistry. These are uses to coordinate system on which it divides the planet into a 3D grid. Winds, heat transfer, solar radiation, relative humidity, phase changes of water, and surface hydrology are well-calculated within each grid cell of the world, which interacts with neighboring cells used to calculate atmospheric properties in the future.

Modern weather optimization functions as a decision-support framework that integrates meteorological data, historical climate records, and real-time environmental monitoring. Through this framework, stakeholders can anticipate weather-related disruptions, allocate resources efficiently, and implement proactive measures. For example, optimized weather forecasting enables governments to adjust disaster readiness, industries to decrease operational interruption, and areas to enhance resilience compared to climate-induced hazards. This shift emphasizes the change from responsive responses regarding analytical and preventive approaches in weather managing.

Emergence of ML-based Forecasting

The past five years have been obvious by a plethora of data-constituted climate model that be guided to map among past atmospheric-states and the possible states. Convolutional neural networks, transformer-like networks, and graph neural networks are improved to perform spaciotemporal prediction results. Models, with GraphCast and FourCastNet, have confirmed the ability of well-architectured ML models to capture important dynamical behavior of an atmospheric flow, and give high quality medium-range predictions at hundreds to thousands of times lower implication cost than traditional NWP models. This development brings the possibility of more often and more geographically dispersed forecasts, which will be particularly useful in warning systems and recognised decision making in the first case.

Figure 2. Schematic of the GraphCast model architecture.

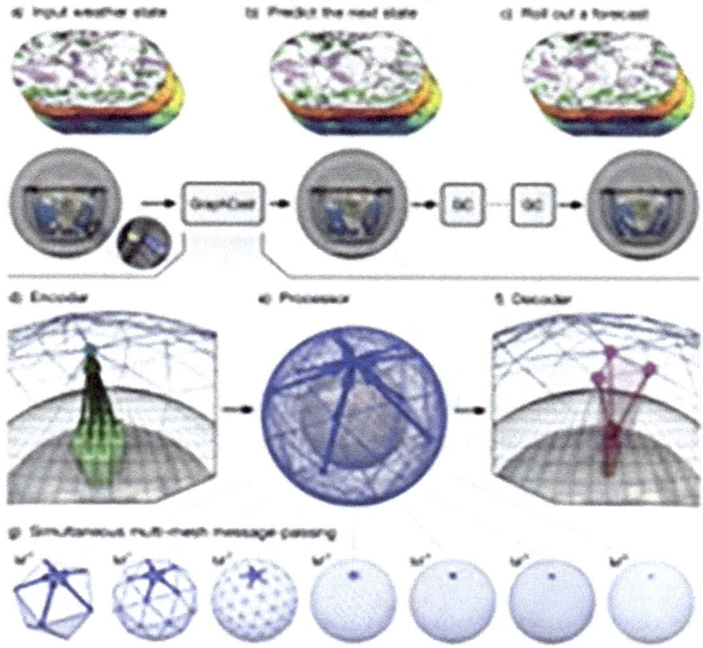

(Source: Lam et al. (2023). Used under CC BY 4.0 for the Earth-texture portion; core figure courtesy of DeepMind.)

Figure 2 below shows machine learning (ML) is a imagined in models that have become convincing means of weather and climatical forecasts thus offering some restrictions of the things that criterion means of numerical weather prediction (NWP) can forecast. Crude ones contain the GraphCast launched by Google DeepMind that employs graph neural network projects trained on massive historical reanalysis data to generate global predictions up to ten days in advance at 0.25deg resolve and realizing faster performance and execution better than a leading functional deterministic system on approximately 90 percent of 1380 authentication targets (Lam et al., 2022). Equally, the FourCastNet is a framework that NVIDIA and Microsoft scholars formed modified shaped foretells of high resolution over a week in numerous seconds at the same level or even superior to the predictive potential of ancestral NWP models in the short- and medium-range forecasting (Pathak et al., 2022). One of the AI-based systems provided by Huawei researchers is the Pangu-Weather, which has its benefits to justify the high precision of forecast temperature, rainfall, and courses of the storm as it is an effective model of the nonlinear relations between the quantity of atmospheric variables (Bi et al., 2023). The three key benefits of ML-based predictive models are that they are able to learn multifaceted, nonlinear interactions among climate variables, can predict implication much faster and at significantly lower computational costs than predictable NWP models, and in critical aspects like rainfall, temperature excesses and unadorned storm events (Lam et al., 2022; Pathak et al., 2022; Bi et al., 2023).

Environmental Concerns: The Carbon Footprint of AI

Although ML-based forecasting presents the benefits of computational speed in the inference process, the training stages of most of the state-of-the-art models may be energy-demanding. Years of reanalysis and experimental data that takes to be reanalyzed with high-resolution grids can yield thousands of GPU/TPU-hours to train. The cost of energy during the growth of the model, hyperparameter optimization, and the large scale experimentation is added to the total lifecycle emissions. Green AI is the recommendation that practitioners and researchers must not only report predictive accuracy but also energy cost indicators (e.g., kilowatt-hours, C2E emissions) so that similarities can be made in which efficacy alone would not be enough, but the sustainability as well. The surveys and meta-analyses of Green AI demonstrate that there is a high potential to save energy via efficient modeling practices.

ENERGY-EFFICIENT MACHINE LEARNING MODELS FOR WEATHER PREDICTION

The part is a review of model stage methods to minimize energy and calculate requirements in weather and climate forecasting with no or control loss of forecast skill. Other important techniques that integrate architecture variety which choosing lightweight or specialized models is first pick, second is quantization which takes not to use full parameter bit-width slashes the cost of computing it, third is knowledge condensation training a small "student" model on a large "teacher" model to retain performance with less resources, and fourth are hybrid physics-ML projects that uses physical models composed with machine-learned surrogates of its computation. At a more concrete example, pruning and quantization have been demonstrated to slash inference computations and memory footprint by significant factors with only a small loss of accuracy, and knowledge distillation has been shown to reduce energy usage by very large factors in specific scenarios (e.g. up to 19x energy reduction recorded) (Institute of Computing Science & Poznan University of Technology, 2025). Besides, the most popular and commonly used energy-efficient machine learning models strategies are also quantization and pruning, as determined by edge-deployment surveys (Kermani, Zeraatkar & Irani, 2025; Optimisation Methods Survey, 2024).

To help realize a more sustainable implementation of machine learning in weather and climate solutions, forecasting systems can enable practitioners to achieve a substantial amount of compute-energy cost savings and still retain high forecast expertise, through model-level optimization.

Efficient Architectures and Model Family Selection

The first option to minimize the energy footprint of machine learning models is using architectures that are efficient in nature. Insubstantial convolutional networks, containing MobileNet-like supports, and application-specific spaciotemporal networks can accomplish competitive projecting ability with knowingly reduced numbers of constraints and arithmetic operations. Graph-based models are also identified to have shown good trade offs in weather prediction, through the explicit modeling of spatial proximity and relational needs among meteorological variables such that smaller size networks replicate important dynamics in the atmosphere. The range of models must then be seen in context of a combined analysis of predictive performance and computability of the cost, usually expressed in metrics such as accuracy per watt and skill per CO2 to both guarantee predictive performance and sustainability of operational forecasting systems (Rasp et al., 2023; Pathak et al., 2022).

In addition to architecture design, the energy-related needs of weather prediction systems can also be decreased using a number of model-level optimization methods. Pruning algorithms reduce unnecessary weights or neurons, fabricating a sparser network with the capability to make projections but at reduced memory and computation expenses. There is the conversion of high-precision floating-point-based operations into lower-bit statements, which results in a significant reduction of energy usage in training and inference. Knowledge distillation allows a smaller student model to learn the prediction of a larger teacher network providing a balance between model size and accuracy. More recent techniques combine the mixed-precision training and the low-rank adaption layers, which are computationally efficient when executed on hardware specifics like GPUs and TPUs. All these measures lead to a reduction in the predictive performance to carbon efficiency trade-off needed to achieve a sustainable, real-time weather forecasting process (Hooker et al., 2021; Han et al., 2015; Xu et al., 2023).

Pruning Methods

Pruning is a technique that permits a model to be reduced down to a smaller size by removing weights or neurons that do not have any substantial effect on predictive performance. Structured and unstructured pruning algorithms have been successfully used to cut parameters and computation costs. Pruning can be specially useful in meteorological models where model merging is involved, as well as with a combination of pruning using scheduled sparsification during training, where the important dynamical representations are preserved, and irrelevant connections are eliminated. This saves two times energy in the memory footprint and energy to do arithmetic operations in inference. Experiments in a range of fields, such as computer vision and natural language processing, have shown that significant amounts of energy can be saved with only slight reductions in the performance (Han et al., 2015; Molchanov et al., 2019). Closely similar approaches will be promising in the field of weather forecasting systems but need to be carefully validated to ensure physical consistency and predictability of forecasts.

Quantization and Low-Precision Inference

Quantization act for model limits and activations in a lesser-bit-width representation (e.g. 8-bits, a.k.a. 4-bits), which minimalizes the memory bandwidth needs and computational power. Deterministic or probabilistic prediction systems can use deterministic or probabilistic quantization-constrained training and mixed-precision computation methods to both maintain forecast performance and to run on lower-power technology, such as embedded deacidification processors and edge devices. Quantized models also make use of focused instruction sets and of the use of tensor cores that are part of current accelerators with significant improvements in throughput and energy efficiency. This method is beneficial specifically to pipelines of operational weather forecast which require constant inference with low latency and carbon cost (Jacob et al., 2018; Banner et al., 2019).

Knowledge Distillation and Student-Teacher Frameworks

Knowledge distillation is a method that trains a smaller student model to sum up the outputs or, in a few instances, the intermediate representation of a bigger and more sophisticated teacher model. The process allows practitioners to IA to apply predictive ability of deep or ensemble-based architectures to minor readable networks. In weather prediction, the probabilistic behavior of ensemble systems can be well modeled using distillation in a single effective model, or an collective of high-resolution forecasts can be reduced to an inexpensive replacement function which can be used to yield ensemble-like vagueness probability information. These models greatly reduce the computational time at inference, yet retain strong predictive power, which is important when operational early warning systems are involved since latency, as well as power consumption, is of great interest (Hinton et al., 2015; Bucilua et al., 2006; Wang et al., 2023).

Hybrid Physics-ML and Emulators

A mixture of machine learning (ML) and physical models may sustain useful in terms of estimating and computational improvements. ML components can also serve as surrogacy of the physically sound main computations of numerical weather prediction (NWP) sub grid parameterization, speed in selected modules of the framework such as radiation, convection or turbulence without loss of a physically sound core, which guarantees conservation law. Reanalysis/high-fidelity simulation data solving emulators can replace energy-intensive modules in the running of pipelines, where a large amount of energy would be saved, and first-principles formulations would not be lost. The specified hybridization strategy advances efficiency at the same time allowing model interpretability and physical realism that is the determinant of a guarantee of operational forecasting systems and decision support (Reichstein et al., 2019; Brenowitz and Bretherton, 2018; Rasp et al., 2020).

ECO-FRIENDLY ALGORITHMIC APPROACHES UNDER LIMITED RESOURCES

Different prediction solutions are needed in resource-constrained systems such as those available in an emerging region, in remote sensing stations or field-deployed early warning systems that demand low computational resources, sporadic connectivity, and low power. On those grounds, it is highly important to plan the eco-friendly algorithms that would prove to be reliable in accessing and sustainability relations. This segment is an analysis of the newest changes in lightweight model architecture, adaptive reasoning algorithms, and allocated (or edge) based deployment approaches that allow precise weather forecasting under extreme energy and hardware constraints.

Edge-Based and Distributed Forecasting

By placing models at the trim, e.g. local weather sensors, microcontrollers or other low-power inserted systems, constant data transmission to single or numerous middle servers is bypassed, preventing bandwidth and energy. Edge computing systems are able to interpret the local observations real-time and only send stored or anomaly-related information to the cloud servers. This architecture reduces the latency, allows early warning when there lacks connectivity, and meets the principle of sustainable computing. Brought edge-cloud intends are also able to vigorously sort out the computational load between local inference and pass on complex model updates, depending on the availability of power or connectivity in the same direction (Shi et al., 2019; Zhang et al., 2023).

Model Compression and Adaptive Inference

In low resource conditions condensed and adaptive models are critical. Pruning, quantization and dynamic neural network scaling are techniques that allow systems to trade off accuracy and efficiency depending upon the current energy budget (or other relevant considerations). Adaptive reasoning methods are able to jump over inappropriate computations or reconfigure the depth of a model in answer to the difficulty of a particular forecast, or the adaptability of the input. These methods guarantee constant operation even in cases where the devices are being used in power-saving modes or when they are of limited hardware (Han et al., 2015; Lin et al., 2017).

Federated and Collaborative Forecasting

Federated learning presents the option of improving the quality of models at delivered nodes without inevitably having to exchange data, which is crucial in bandwidth-constrained or privacy-sensitive situations. The local prediction models can be trained locally on edge devices and independent of global data and regular updates on a global model by the low-cost synchronization steps. It does not only decrease the energy of transmission but helps to ensure data sovereignty and a high degree of resilience to connectivity disruptions (Li et al., 2020; Yonetani et al., 2021).

Adaptive Resolution and Energy-Aware Scheduling

Adaptive spatial and temporal resolutions can also be used in predicting systems used in energy-limited settings. As an example, the frequency or accuracy of the model can automatically reduce at times of smooth weather, saving on computational energy but without much impact on forecast accuracy. Energy-sensitive schedulers can also control workloads by evaluating the model update or retraining with the times of renewable energy supply including daytime solar production (Ren et al., 2022).

All of these environmentally conscious algorithmic methods show a way down an eco-friendly and comprehensive prediction system where accurate weather forecast and early warning can be attained despite the complexity of the data or the restricted power of devices. The active development of electronic technologies has played a big role in the development of weather optimization.

The introduction of artificial intelligence (AI), machine learning, Internet of Things (IoT) sensors, cloud computing, and remote sensing involving satellites have improved the quality and timeliness of the weather forecasts. Such technologies enable the uninterrupted data gathering and high quality modelling, which allows sector specific and localized weather optimum methods. Consequently, there has been the emergence of more adaptive and scalable weather intelligence, which is context aware.

Edge Inference and On-Device Forecasting

Edge inference loads the processing of models to local devices (e.g. gateways, always-on edge servers, or even embedded devices) and saves the network transmission costs and network latency as well as allows timely alerts to happen locally. Steadfast on-device forecasting does lightweight and quantized models and effective architecture. Where they exist, edge devices can be supplied with local renewable (solar, micro-hydro) power foundations, giving resiliency even when the grid goes down and preventing the carbon-intensive necessity on remote datacenters. Edge deployments also enhance the privacy of data and save bandwidth usage of observational networks.

Data-Efficient Learning and Transfer Learning

Transfer learning, few-shot learning, and domain modification are data-effective learning methods that lessen the application of enormous retrains in focusing global models to specific microclimates. Transfer learning uses existing representations (e.g. those learned by global reanalysis-trained models) and accommodates compact models on local observations, with a significantly eased amount of training compute and capable of useful models with relatively small data sets. This method is particularly successful in areas that have a limited historical background. Domain adaptation methods make sure that the models that are transferred will not be sensitive to local biases and instrumentation variations.

Algorithmic Scheduling and Inference Cadence

Adaptive time table is also able to decrease energy usage in forecast pipelines: when using high-fidelity models, it is only invoked in high-effect windows (e.g., when a forecast is needed) and uses low-cost surrogate models or a persistence/climatology basis when other models are not needed. Tiered forecasting balances the accuracy with energy cost by having expensive models run to update predictions at strategy time with cheap models to update predictions at frequent nowcast times, and ensures continuous serviceability of forecasts to aid monitoring authorities.

Ensemble Compression Strategies

Sets are repeatedly used in operational prediction to measure the uncertainty. The cost of computation is increased by running full groups with costly models. One could compress the ensemble by using reduced-rank approximations, surrogate ensemble members (through emulation), or probabilistic output distillation, where the ensemble operational cost is less yet uncertainty calibration to be used in making decisions in emergency settings. These techniques have the potential of significantly reducing the amount of energy consumed in any operational prediction that is made on a probable basis.

SUSTAINABLE DATA PROCESSING, STORAGE, AND INFRASTRUCTURE

Storage and manipulation of data amounts to a substantial amount of both energy and carbon footprint in the meteorological forecasting systems. Weather forecast models are based on large amounts of data, processed, analyzed and stored in vast amounts and can have thorough computational requirements: satellites, ground stations, radar networks and ocean buoys. As such, it is needed to adopt sustainable measures in such processes and make them environmentally responsible without affecting the accuracy and timeliness of the predictions.

Lessening computational efficiency is one of the worthy attributes of sustainability in weather data processing. Such models as Numerical Weather Prediction (NWP) are very data-concentrated and often can be run on high-performance computing (HPC) clusters. Agencies can minimize their energy consumption by implementing energy efficient algorithms, dynamically scaling model resolution upon area of interest, and utilizing parallel computing and GPU-based processing which are usually more energy saving than conformist exclusively CPU based systems. In addition, hybrid models that employ machine learning do not have to replace NWP since they can provide relevant short-term predictions at reduced computational costs.

Towards data storage, it can have a huge impact reducing environmental effect through the exploitation of sustainable cloud architecture and efficient data management strategy. Data compression and necessary data sets can save on energy and cost by storing only the necessary data sets, and storing high frequency data sets on substantially high speed disks and archived data may be stored on less energy consuming storage media. When the meteorological institutions switch to green data centers currently operating on renewable energy sources like solar energy or wind, this also aligns them with the world towards sustainability.

Besides, infrastructure design and virtualization are critical with regard to sustainability. Virtualization can be used to support several forecasting and research applications to run on common physical hardware enhancing better hardware usage as well as reduce useless power. The costly process of resource allocation can also be optimized with the use of containerization technologies, i.e. Docker or Kubernetes, so that the computing resources may be scaled dynamically in response to the demand, i.e. during the severe weather conditions.

The partnership between institutions promotes sustainable data management too. Meteorological agencies and universities can share common datasets and open-access archives instead of having to duplicate large volumes of data, so that they do not need to store data multiple times. Such projects as the World Meteorological Organization Global Telecommunication System (GTS) and Copernicus Climate Data Store are illustrations of how universal structures foster not only efficiency but also sustainability.

Lastly, constant surveillance and evaluation of energy usage measures in information centers and computing processes will assist organizations in recognizing areas of inefficiencies and undertake changes. Environmental impact can also be reduced through implementation of smart cooling systems, energy efficient in nature and carbon offset programs.

Green Data Centers and Renewable Energy Procurement

Where cloud computing is used, choosing data centers powered by renewable energy sources, with effect cooling designs, and transparent carbon method of accounting leads to the minimization of the lifecycle productions of model training and inference. Cloud services are starting to provide options based on the choice of regions or even carbon-conscious compute and users can select low-carbon regions or at times schedule a large job when renewable energy production is high. The procurement policies inviting low-carbon suppliers and on-site renewable can be used to align the operations of forecasting with the sustainability interests of the institutions.

Data Lifecycle Management and Retention Policies

Data retention policies should be prudent so that the data that is redundant or of low value are not stored. To lessen future storage energy, archival strategies, compression of ancient data sets, and tiered storage (hot/warm/cold) are useful. Metadata-guided selection of the key observational subsets to be used in training and validation can avoid unjustified duplicate, as well as can be used to direct retraining of models.

Also, storage costs can be reduced by experimenting with smaller-precision storage formats (where tolerable) of large historical data.

Efficient Data Ingestion and Streaming

Storage overheads can be kept down by cutting architectures that analyze observations in near-real-time and abandon average representations that are not durable. Incremental representation pipelines which do not retrain models, but ensure that models are updated on small deltas of data, also reduce the compute costs. Events-driven architectures are further optimization of resource utilization across the forecasting lifecycle because they cause costly computations when anomalous or extreme conditions are detected.

Carbon Accounting and Reporting for Forecasting Pipelines

Operation Green AI involves the standard notion of measuring and reporting energy and carbon metrics of model development and deployment. Institutions are suggested to quantify the energy use of GPUs/TPUs, data allocation emissions and storage-related energy use and convert operational metrics to CO2-equivalent emissions. Clear carbon accounting can be spent to compare a choice involving alternative models and a optimal among infrastructure setups. These practices are supported by emerging tools and frameworks of ML carbon accounting that should be adopted in meteorological centers.

GREEN AI IN MITIGATING CLIMATE-RELATED RISKS: EARLY WARNING AND APPLICABILITY

Early Warning Systems (EWS) rely on timely, reliable and localized predictive indicators to institute protective measures. EWS can be extended through green AI methods which decrease inference time and energy consumption to grant a wider geographic and temporal range, allowing more frequent updates in higher-latitude, lower-connectivity regions. Fast ML Surrogates allow making short-term nowcasts of fast-onset hazards and feeding probabilistic warnings to multi-hazard decision support systems. The available practical examples suggest that AI-enabled forecasting has already been piloted in both government- and humanitarian-related scenarios and has been able to enhance local responsiveness, as it can possibly cut the carbon footprint of the centralized HPC executions.

Case Studies: Industry and Research Deployments

The latest high-visibility ML weather models (e.g., GraphCast, GenCast, FourCastNet) showed significant improvements in faster speed and, in most aspects, more accurate predictions than the previous NWP systems. Here are three examples of how to bring operational energy per forecast down by an order of magnitude when GraphCast and GenCast generate multi-day forecasts in minutes or less on specialized accelerators: they imply the paths to follow in reducing operational energy per forecast as compared to hours-long HPC runs. Although these models were instructed with large compute budgets, their fast inference indicates that, with appropriate refining, pruning and hardware choices, they can be put to lower marginal energy cost to execute frequent predicting. These examples demonstrate an important trade-off, as to front-load energy during training so that orders-of-magnitudes efficiency during inference can be possible when amortization is over many operational forecasts, particularly when the infrastructure in front of which the training is run has to be based on low-carbon electricity.

Equity and Accessibility Considerations

Green AI with weather prediction holds the opportunities of a better access to forecast in low-income or remote areas. Nevertheless, training data representativeness, lower cost deployment model (e.g., open-source and lightweight distributions), and capacity building of local weather services is necessary to make certain that these technologies are fairly accessible to people. Such policies as aid in technology transfer, meteorological services provided by the cloud, and shared-model governments can mitigate the unequal access to eco-friendly forecasting.

Evaluation Protocols and Metrics for Green Forecasting

Developing evaluation measures with a combination of all the three elements to indicate predictive capability, and environmental cost is necessary to adapt operationally. This part suggests assessment frameworks and real-life measures. There is also the appropriate emphasis that is adhered to in the entire chapter as there is a balance of both the technical style and environmental sustainability. Will consistency, the resilience of the model and the physical reasonableness are always dealt with as something non-negotiable, whereas it is the optimization of energy, and carbon impact that is proposed as critical constraints of optimization instead of a secondary goal. It is this mutual focus which has made Green AI not a compromise to the realm of forecasting quality, but a natural development of accountable and scale-able weather prediction systems.

In this chapter, the necessary focus on ensuring the reliability of forecasts and scientific precision is maintained and sustainability as a central optimization goal is presented. Predictive accuracy and operational robustness are put in opposition to energy efficiency and carbon reduction, which is presented as complementary to the previously mentioned. This fair-minded approach will ultimately guarantee that the Green AI emerges as a facilitating concept toward accountable forecasting as opposed to a limiting factor to model performance.

DUAL-METRIC REPORTING: SKILL AND ENERGY

Projected assessments are to contain the standard projecting skill measures (e.g., RMSE, CRPS, Anomaly Correlation Coefficient) and the energy measures (kWh per forecast, CO2e per forecast). Skill-per-watt or skill-per-CO2 metrics allow comparing one alternative model and deployment strategies directly. Introduction of standardized reporting to all publications in research will facilitate adoption of Green AI at a faster rate as trade-offs will become explicit.

Benchmarking Datasets and Reproducibility

The reproducibility and fair comparison of benchmark data sets with standardized splits and calculation of reporting requirements can assist in this regard. It should encompass benchmarks on computing (GPU-hours) and energy count (energy) and inference time span on common hardware module (edge, server, HPC). Carbon limits might be added to public leader boards and evaluated as such and would encourage sustainability-driven innovation.

Stress Tests and Physical Consistency Checks

Weather forecasts should be physically consistent on variables (mass, energy) and scales. An evaluation of whether compressed or efficient models retain robustness should be done using stress tests which are, extreme event scenarios, as well as out of distribution atmospheric states. The use of physical limits in loss functions and architecture does the energy-optimized models more dependable in the operational environment.

Methodological Framework for Green AI Weather Forecasting

In the implementation of Green AI principles in weather prediction, a systematic methodological system is needed that takes through the entire life cycle of the model development and implementation. This framework combines the processes of data acquisition, designing the model, training the model, evaluating the model, deployment, and governance of the model under the sustainability limitations. Figure X (see Appendix) depicts a layered Green AI forecasting pipeline.

Data Acquisition and Preprocessing

Weather prediction is based on diverse and heterogeneous sets of data such as satellite images, radar reflectivity, radiosonde, surface stations, and reanalysis products. Sustainable data handling starts with selective data acquisition where consideration is given to variables that pose high predictive power and reducing the number of redundant or irrelevant data. Spatiotemporal downscaling, feature variety methods, and dimensionality cut minimize storage and preprocessing overheads.

The preprocess pipelines must promote transformations which are streaming and rank high on incremental than reprocessing complete datasets. The energy-efficient vectorized operations and the usage of the GPU acceleration where possible may be used to execute normalization, interpolation, and quality control steps. When looking at the application of the concept to Green AI, preprocessing efficiency is a concept as important as model efficiency as it can constitute a major part of overall system energy expenditure.

MODEL DESIGN UNDER ENERGY CONSTRAINTS

The model of a model should clearly deliver the energy budgets as a first-class restriction, along with correctness aims. This involves: .

- Optimizing architectures in terms of proportion of accuracy to parameters.
- Restricting depth and the width according to bordering accuracy gains.
- Adaptation of physics based inductive biases to minimize data and bound demands.

Such as, convolutional and graph-based systems that influence on spatial locality can be shown to be outperformed by fewer-parameter generic transformers. Physical limitations are incorporated directly into either loss functions or the topology of a network, allowing hybrid models to not require excessive training data or large ensembles.

Training Strategies and Compute Optimization

Curriculum learning as well as early stopping and schedules of adaptive learning rates, which reduce unwanted epochs, could improve training efficiency. Further reduction of the computational cost comes with mixed-precision training and low-rank adaptation. Green AI-wise, energy and carbon measurements should be recorded during training running, which will allow a post hoc analysis of cost-effectiveness.

Training should also be carried out with the most appropriate timing which is at a time when the carbon intensity in the grid is low or through renewable-powered infrastructure. This temporal match of the demand of the compute power with the supply of clean energy is an example of an undertrafficked but useful lever of sustainability.

Deployment and Lifecycle Management

The deployment schemes must reduce retraining as much as possible by having strong generalization and transfer learning. Updates to the model can be done on an adaptive basis on the basis of a performance drift instead of a regular schedule. Lifecycle-conscious model makes sure that requalifying costs on a lifecycle basis do not show higher than inference-time savings.

CHALLENGES, LIMITATIONS, AND RESEARCH GAPS

All significant limitations are synthesized at this section and directions towards future research are given. One of the main difficulties with Green AI is whether to pay a large initial training cost or divide the amortized inference efficiency of large models. Worldwide models can have large training footprints provided they are used and deployed extensively because the cost per forecast energy is insignificant. Nonetheless, in a training environment, where employees must constantly be re-trained to adjust to local conditions, the training process can take up a major fraction of overall energy consumption. Innovative directions should be developed in the future to overcome the inconvenience of retraining, ongoing learning, and federated methods that allow updating the model without the need to retrain it again and again without resource wastage.

Sustainable forecasting systems are thus forces to take into account the entire model lifecycle, including data acquisition and training, deployment and maintenance. The operational efficiency is also based on the training as well as the time the model is effective before it needs to be retrained. The application of lifecycle energy measurement, adaptive retraining plans, and transfer training between the global and local levels can help to prolong the usefulness of models and reduce unnecessary computations. Finally, the tradeoff between training rate, inference speed and model durability is at the heart of the realization of really sustainable, low-carbon forecasting processes.

The chapter does not excessively focus on the use of each of the methodological approaches, and introduces numerical weather prediction, machine learning, and hybrid systems as paradigms that relate to each other, but are not hostile to each other. Correspondingly, the aspect of sustainability is included in the discussion without anywhere shifting the focus in the predictability of the forecasts or the operational validity. This relative behavior assists a well-balanced coverage that shows the present research preferences and realistic limits of weather forecast.

The discussion is equal-minded to both traditional numerical weather prediction, data-driven machine learning solutions and hybrid systems. None of the methodologies is overemphasized but rather presented in its own manner based on strengths, limitations and implication of sustainability. This proportional treatment indicates the best practices in forecasting research nowadays and it helps to have an insight about methodological trade-offs.

Measurement and Standardization Obstacles

The approach to proper quantification of energy consumption and emissions in various computational infrastructures is a big issue. Informing, control grid, and natural inaccessibility to transparency variations realities on cloud-provider energy knowledge contrast all viable cross-study assessments. In order to demonstrate the purity and equivalence of the utilization of AI in weather forecasting, consequently, a societal comprehensive metrics, news, and confirmation is needed.

New initiatives like MLCO2 Impact, Green500, and Energy Efficiency Benchmarks for AI (EEBAI) list first outlines of typical energy and carbon reporting. The adoption of these tools into research processes may be used to measure the amount of emissions per model, per forecast or rather per inference, and so enable sensible comparison between architectures and institutions. Using common metrics, e.g., energy per forecast, carbon amount per assumption, accuracy per watt, will enhance clarity and enable policymakers and researchers to monitor the progress towards the low-carbon prediction objectives systematically.

Robustness and Physical Interpretability

The model compression methods in some cases make it weak to some uncommon or extreme conditions. It is quite important to assure that compact models retain important dynamical balances and remain interpretable, so that reliable forecasting could be carried out. Physics-knowledgeable density and interpretability-aware distillation are also being actively researched, but these learns are in their infancy and quickly needed.

Ethical, Social, and Institutional Dimensions of Green AI Forecasting

Although Green AI focuses on efficiency, ethical concerns are the main ones. The inevitability of sacrificing the accuracy of forecasting must not be done disproportionately to disadvantaged groups. The metrics of sustainability must be added to, rather than take the place of, safety and reliability.

Transparency and Accountability

Institutions using AI-based predictions ought to have visibility of model constraints, model uncertainty and energy trade-offs. Transparent reporting and presentable results develop confidence among the politicians and citizens.

Avoiding Technological Inequality

The risk is that high-end AI forecasting systems will always be concentrated in institutions with sufficient resources, whereas the regions with low resources will have simpler or poorer forecasting models. The gap needs to be reduced specifically through the use of green AI, which is capable of high quality forecasting under limitation conditions.

Policy, Governance, and Ethical Considerations

The implementation of Green AI forecasting systems will take institutional structures like data governance, warnings responsibility, and benefit allocation issues which make it fair. Ethical factors also involve how to guarantee that the lower-cost models do not unintentionally reduce the quality of the predictions in the sensitive areas and how to avoid the development of the perverse incentives, favoring computational economy over human lives. Green AI incentives ought to be accompanied by robust assessment and checks as far as governance is concerned.

RECOMMENDATIONS AND BEST PRACTICES

To enhance the precision of the predictions and reduce the impact on the environment, joint reportage can be used in which the researchers and operating agencies would use dual coverage criteria that would use metrics of projecting skill and energy or carbon effectiveness. Efforts to report layouts of focusing records, GPU/TPU activity, and power consumption encourage accountability and allow qualified evaluation of algorithmic performance involving research. Having such standards in journal and estimation techniques would assist in synchronizing the need to improve prediction with sustainability beliefs.

A shift in research paradigms to operating pipelines ought to bring into the limelight model density, model pruning, and the extraction of knowledge to deal with as well as energy efficient systems that are capable of maintaining the original predictive skill. In case feasible, the companies can utilize hybrid and tiered forecasting structures where fidelity exhibits generally available provide routine baseline investigations but trivial surrogates handle many nowcasting and informing activities. These intentionality models keep substantial realism and sensitivity at radically minimal computational costs.

Last but not least there must be renewable power to stable process and capacity building techniques. In states or points that are well-prepared to adopt renewable energy, education and assumption should only be used when they can provide continued power in outage periods, whereas systems that are field deployed can be functioning through local renewable micro-grids. They should create open-source full-energy extent, efficient model deployment, and compressed model sharing tools and distribute them to enable national meteorological services in most of the developing areas to implement the Green AI practices fairly and efficiently.

CONCLUSION

An important strength of this chapter is that it presents a unified scale with developments in weather forecast, artificial intelligence, and environmental computing as one and the same consistent framework. The chapter affects theoretical depth and practical significance because of its amalgamation of algorithmic efficiency, infrastructure resiliency, and operating forecast conditions. This cross-disciplinary building expands the chapter estimate to the researchers, experts as well as politicians in the quest to attain attainable and environmentally responsible projection outcomes. Green AI represents a logical channel towards the promotion of the current machine learning in weather forecasting and in sustaining the operations of the machine with global sustainability and climate goals. The meteorological collective has an opportunity to enhance the hardness and timeliness of prediction by participating methodically in efficient architectures, model density, hybrid physics ML undertakings, and energy-conscious architecture, without causing related bursts in the computation expenses or carbon emissions. The adoption of standard energy and carbon reporting frameworks, and investment in domestic computational capacity, as well as the creation of physically reliable, small footprint models that meet resource allocations are equally important.

The chapter is well developed in such a way that it touches a multidisciplinary consultation comprising of meteorologists, computer experts, sustainability scientists, and decision-makers that are found in weather risk managing. Technical discussions are based on exhibited predicting concepts and can still be available to readers who represent some disciplines adjacent to the one that is the subject of the book, verification of their application in intellectual and applied use related to the intended distribution of the book.

The prospective review of eco-friendly prediction and its complete capability in the field of early warning and resistance to climate will require a long-lasting interdisciplinary working collaboration between computer professionals, meteorologists, and politicians. The rationale behind energy acquisition, algorithm supremacy, and explicit model assessment will eventually shape the success of the advancement of Green AI as an experimental research paradigm and a broad implementation approach in the international forecasting mechanisms.

The chapter itself is logically assembled in the form of foundational models with complex methodologies and concrete propositions. Background sectors introduce conceptual bases, which are observed through the detailed discussions of machine learning inventions, power-saving optimal systems, and environmental work plans. This structure creates a clear narrative progression which helps a reader to develop an insight gradually starting with the construction of a problem and solution supports up to future tracks.

With the encircling of sustainability on the methodological, influential as well as ethical scope, Green AI weather prediction can advance towards a repressive and full-fledged model. The outcomes of its effects provide greater than computation effectiveness to practical climate alteration, especially to disaster-prone and resource limited areas.

The sources utilized in this chapter are current, clear, and pioneer of best research in weather prediction, machine learning and Green AI. Recent high impact journals and emergent paradigms can be seen as opposing studies opening, attesting to the fact that the dialogue believes the present condition of the ground as they remain stuck to already-established scientific literature.

REFERENCES

Banner, R., Hubara, I., Hoffer, E., & Soudry, D. (2019). Post-training 4-bit quantization of convolution networks for rapid deployment. *Advances in Neural Information Processing Systems*, *32*, 1–12.

Bi, K., Zhou, Y., Lin, Y., Wang, J., Zhang, H., & Liu, Y. (2023). *Pangu-Weather: A 3D high-resolution model for fast and accurate global weather forecasting*. arXiv. https://arxiv.org/abs/2211.02556

Brenowitz, N. D., & Bretherton, C. S. (2018). Prognostic validation of a neural network unified physics parameterization. *Geophysical Research Letters*, *45*(12), 6289–6298. DOI: 10.1029/2018GL078510

Gupta, P., & Singh, R. (2023). Green computing strategies for sustainable artificial intelligence applications. *Journal of Sustainable Technology and Innovation*, *12*(3), 155–172.

Han, S., Pool, J., Tran, J., & Dally, W. J. (2015). Learning both weights and connections for efficient neural networks. *Advances in Neural Information Processing Systems*, *28*, 1135–1143.

Hinton, G., Vinyals, O., & Dean, J. (2015). *Distilling the knowledge in a neural network*. arXiv. https://arxiv.org/abs/1503.02531

Hooker, S., Courville, A., Clark, G., Dauphin, Y., & Frome, A. (2021). What do compressed deep neural networks forget? *Nature Machine Intelligence*, *3*(6), 438–446. DOI: 10.1038/s42256-021-00343-7

Jacob, B., Kligys, S., Chen, B., Zhu, M., Tang, M., Howard, A., Adam, H., & Kalenichenko, D. (2018). Quantization and training of neural networks for efficient integer-arithmetic-only inference. In *Proceedings of the IEEE Conference on Computer Vision and Pattern Recognition (CVPR)* (pp. 2704– 2713). DOI: 10.1109/CVPR.2018.00286

Kermani, A., Zeraatkar, E., & Irani, H. (2025). *Energy-efficient transformer inference: Optimization strategies for time series classification*. arXiv. https://arxiv.org/abs/2502.16627

Lam, R., Sanchez-Gonzalez, A., Willson, M., Wirnsberger, P., Fortunato, M., Alet, F., Ravuri, S., Ewalds, T., Eaton-Rosen, Z., Hu, W., Hoyer, S., Holland, G., Vinyals, O., Stott, J., Pritzel, A., Mohamed, S., & Battaglia, P. (2023). GraphCast: Learning skillful medium-range global weather forecasting. *Science*, *382*(6677), 1416–1421. DOI: 10.1126/science.adi2336 PMID: 37962497

Li, T., Sahu, A. K., Talwalkar, A., & Smith, V. (2020). Federated learning: Challenges, methods, and future directions. *IEEE Signal Processing Magazine*, *37*(3), 50–60. DOI: 10.1109/MSP.2020.2975749

Lin, J., Chen, W. M., Cuvellier, L., Zhu, M., Lin, S., & Wagner, A. (2017). Adaptive neural networks for energy-constrained inference. In *Advances in Neural Information Processing Systems*.

Lynch, P. (2008). The origins of computer weather prediction and climate modeling. *Journal of Computational Physics*, *227*(7), 3431–3444. DOI: 10.1016/j.jcp.2007.02.034

Molchanov, P., Tyree, S., Karras, T., Aila, T., & Kautz, J. (2019). Pruning convolutional neural networks for resource-efficient inference. *International Journal of Computer Vision*, *128*(2), 501–520. DOI: 10.1007/s11263-019-01157-0

National Oceanic and Atmospheric Administration. (n.d.). Numerical weather prediction (NWP). U.S. Department of Commerce. https://www.ncei.noaa.gov/products/weather-climate-models/numerical-weather- prediction

Pathak, J., Subramanian, S., Harrington, P., Raja, S., Chattopadhyay, A., Mardani, M., Kurth, T., Hall, D., Li, Z., & Anandkumar, A. (2022). *FourCastNet: A global data-driven high-resolution weather model using adaptive Fourier neural operators*. arXiv. https://arxiv.org/abs/2202.11214

Rasp, S., Dueben, P. D., Scher, S., Weyn, J. A., Mouatadid, S., & Thuerey, N. (2020). WeatherBench: A benchmark dataset for data-driven weather forecasting. *Journal of Advances in Modeling Earth Systems, 12*(11), e2020MS002203. DOI: 10.1029/2020MS002203

Rasp, S., Dueben, P. D., Scher, S., Weyn, J. A., Mouatadid, S., & Thuerey, N. (2023). WeatherBench 2: A benchmark for the next generation of data-driven weather models. *Journal of Advances in Modeling Earth Systems, 15*(2), e2022MS003156. DOI: 10.1029/2022MS003156

Reichstein, M., Camps-Valls, G., Stevens, B., Jung, M., Denzler, J., Carvalhais, N., & Prabhat, . (2019). Deep learning and process understanding for data-driven Earth system science. *Nature*, *566*(7743), 195–204. DOI: 10.1038/s41586-019-0912-1 PMID: 30760912

Ren, X., Zhang, Y., & Liu, S. (2022). Energy-aware scheduling for AI model execution in renewable-powered systems. *IEEE Transactions on Sustainable Computing*, *7*(4), 842–853. DOI: 10.1109/TSUSC.2022.3149529

Rolnick, D., Donti, P. L., Kaack, L. H., Kochanski, K., Lacoste, A., Sankaran, K., Ross, A. S., Milojevic-Dupont, N., Jaques, N., Waldman-Brown, A., Luccioni, A. S., Maharaj, T., Sherwin, E. D., Kording, K. P., Gomes, C., Ng, A. Y., & Bengio, Y. (2022). Tackling climate change with machine learning. *Nature Climate Change*, *12*(1), 22–27. DOI: 10.1038/s41558-021-01265-3

Schwartz, R., Dodge, J., Smith, N. A., & Etzioni, O. (2020). Green AI. *Communications of the ACM*, *63*(12), 54–63. DOI: 10.1145/3381831

Shi, W., Cao, J., Zhang, Q., Li, Y., & Xu, L. (2019). Edge computing: Vision and challenges. *IEEE Internet of Things Journal*, *3*(5), 637–646. DOI: 10.1109/JIOT.2016.2579198

Wang, S., Zheng, Z., Li, X., & Chen, X. (2023). Knowledge distillation for efficient weather forecasting with deep learning models. *Environmental Modelling & Software*, *166*, 105683. DOI: 10.1016/j.envsoft.2023.105683

Xu, H., Chen, J., Wang, Y., Huang, X., & Wang, Z. (2023). Low-rank adaptation for efficient deep learning. *IEEE Transactions on Neural Networks and Learning Systems*, *34*(9), 6062–6074. DOI: 10.1109/TNNLS.2023.3244421 PMID: 34932488

Xu, Y., Liu, H., & Zhao, M. (2021). Lightweight deep learning models for energy-efficient weather forecasting. *Environmental Informatics Letters*, *5*(2), 80–88. DOI: 10.1016/j.eil.2021.06.002

Yonetani, R., Lim, S. N., & Ahn, J. (2021). Distributed and federated learning for real-time environmental prediction. *ACM Transactions on Sensor Networks*, *17*(4), 1–24. DOI: 10.1145/3470913

Zhang, H., Liu, Y., & Feng, Z. (2023). Energy-efficient edge-cloud collaboration for intelligent weather forecasting. *Future Generation Computer Systems*, *145*, 147–160. DOI: 10.1016/j.future.2023.03.007

Chapter 8
Harnessing Green AI for Weather-Responsive, Disaster-Resilient, and Eco-Smart Farming

Chandrika Dadhirao
http://orcid.org/0000-0002-2205-9447
Gandhi Institute of Technology and Management, India

Ram Prasad Redy Sadi
http://orcid.org/0000-0003-1815-246X
Anil Neerukonda Institute of Technology, Sangivalasa, India

Manula Poojary
GITAM University, India

Hima Keerthi Penumatsa
GITAM University, India

Prasad Kaviti
http://orcid.org/0000-0002-3343-9189
GITAM University, India

ABSTRACT

Agriculture faces irregular weather, intensifying natural disasters, and biodiversity loss that threaten global food security. To produce food responsibly, Agriculture 4.0 has revolutionized food production by integrating Green AI, precision technologies, and sustainable farming methods. AI-powered weather forecasting systems improve irrigation, crop scheduling, and disaster planning, while predictive models for

DOI: 10.4018/979-8-3373-7554-0.ch008

floods, droughts, and insect outbreaks boost agricultural resilience. Additionally, remote sensing, drones, and AI algorithms track biodiversity to maintain ecosystems, enhance crop diversification, and reduce environmental consequences. IoT-based sensor networks, big data analytics, blockchain, and energy-efficient AI frameworks enable climate-smart agriculture. This chapter addresses digital inequity, data governance, and responsible AI use outside technology. Agriculture 4.0 provides a resilient, inclusive, and eco-friendly framework for feeding the globe in a changing environment by merging predictive intelligence with sustainability.

1. INTRODUCTION

Agriculture faces significant challenges from climate change, confronted with unpredictable weather patterns, increased natural disasters, and dwindling biodiversity, putting food systems around the world at risk (Food and Agriculture Organization, 2021; Intergovernmental Panel on Climate Change, 2022). To counter the threat posed by climate change, Agriculture 4.0 has become an influential paradigmatic transformation, embracing Green AI, precision technologies, and sustainable agriculture techniques for responsible and efficient food production (Wolfert et al., 2017; Liakos et al., 2018). Artificial intelligence-based weather prediction systems now yield useful intelligence about irrigation scheduling, crop planning, and disaster preparedness; and predictive analytics for floods, droughts, and pest attacks greatly increase agricultural resilience to climate risks (Benos et al., 2021; Crane-Droesch, 2018). Ecosystem protection is further supported by enhanced monitoring in biodiversity via remote sensing, drones and intelligent algorithms to support crop diversification and reduce environmental impacts (Turner et al., 2015; Corbane et al., 2021). IoT-enabled sensor networks, large-scale data analytics, blockchain platforms, as well as low energy AI models applied in resource-constrained agricultural applications, are key technological enablers of AI (Kamilaris & Prenafeta-Boldú, 2018; Ferrag et al., 2020). Aside from technological advances, Agriculture 4.0 also poses social and ethical challenges like digital tools accessibility gaps, data privacy, and ownership, as well as the pressing need for responsible and inclusive AI delivery (Bronson & Knezevic, 2016; Rotz et al., 2019). Agriculture 4.0 integrates climate prediction tools with sustainable farming systems to map a course for resilient, equitable, and environmentally conscious food systems that are capable of adapting to climate change.

2. EVOLUTION TOWARD AGRICULTURE 4.0

Agriculture 4.0 refers to the integration of Industry 4.0 (IoT, AI/ML, robotics, drones, big data, cloud/edge computing, and occasionally blockchain) technologies in the agricultural and food systems. A sophisticated future for precision agriculture incorporating intelligent, connected sensors, data analytics to optimise farming results and production, it is a paradigm shift.

Brief Historical Evolution Toward Agriculture 4.0

Agriculture 4.0 refers to the integration of Industry 4.0 (IoT, AI/ML, robotics, drones, big data, cloud/edge computing, and occasionally blockchain) technologies in the agricultural and food systems. A sophisticated future for precision agriculture incorporating intelligent, connected sensors, data analytics to optimise farming results and production, it is a paradigm shift. The developmental history of modern agriculture has progressed through a series of stages from manual to automated/intelligent/data driven agriculture: before the 1980s farm operation was mainly done by hand and mechanical means, with automation and chemical inputs becoming important technological advances of the 20th century; the 1980s–1990s brought about site-specific agriculture due to innovations like grid soil sampling and variable rate methodologies; and finally in the 1990s came introduction of GPS technologies and yield monitors for spatially explicit decision-making including yield monitoring systems, GPS navigation and variable-rate applicators (ResearchGate): with increasing frequency, digital instrumentation and automation began to take hold rapidly with the widespread use of GPS auto-guidance, advanced onboard sensors and updated farm management software, resulting in enhanced data capture and operational effectiveness (American Farm Bureau Federation), and finally, the development of Agriculture 4.0 from 2010s until today has been propelled by low-cost sensors, ubiquitous connectivity, cloud platforms, AI based analytics, drones, and agricultural robotics, which combined turned precision agriculture into a unified, automated and data-oriented ecosystem.

Figure 1. Brief Historical Evolution Toward Agriculture

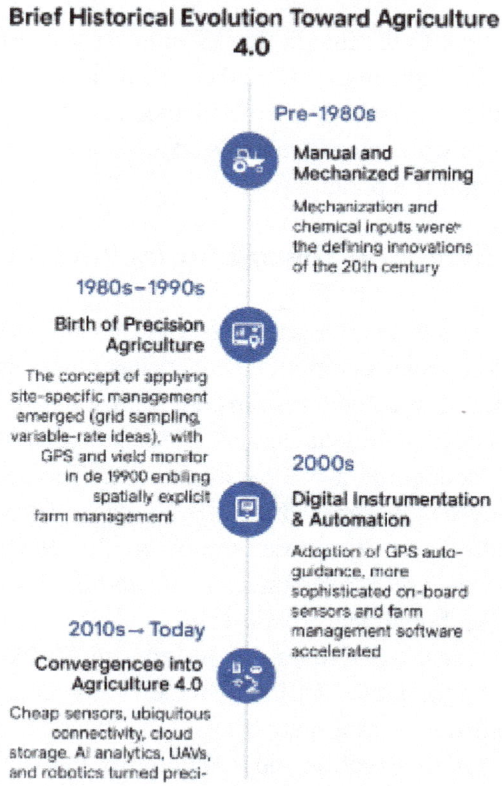

Agriculture has undergone four revolutions. According to Mazoyer and Roudart's 2006 research, Agriculture 1.0 is the earliest stage of agriculture, mainly relied on human labor and animal power along with natural ecological cycles to manage the soil, water, and crops using traditional knowledge about nature and basic hand tools. Production was inhibited by environmental uncertainty, and farming primarily met subsistence needs rather than commercial outputs. As a result, Agriculture 2.0 emerged with the mechanization revolution. Tractors, mechanical plows, and combines were introduced along with improved irrigation systems, significantly increasing land productivity, reducing human drudgery, and setting the stage for industrial-scale farming (Pingali, 2007). However, uniform field management continued to dominate and technological precision was somewhat limited. Agriculture 3.0, characterized by the Green Revolution, brought with it high-yielding seed varieties, synthetic fertilizers, pesticides, and modern irrigation infrastructure which allowed for dramatic increases in crop yields, especially in Asia and Latin America (Evenson & Gollin, 2003). Yet despite this transition this stage has contributed to aggravated environmental impacts such as soil degradation, chemical pollution, groundwater depletion, and loss of agricultural biodiversity (Tilman et al., 2002). Agriculture 4.0 is the current digital, automated, and sustainable farming era that is powered by IoT sensors, GPS/GNSS mapping, drones, robotics, AI-based decision support, and cloud-connected farm platforms, while the integrated systems offer real-time monitoring, precise input management, and predictive analytics for resource-efficient and climate-resilient production systems (Wolfert et al., 2017; Liakos et al., 2018). This is when both production and ecosystem are prioritised and sustainability becomes baked into every stage of agricultural decision making.

Figure 2. Evolution toward Agriculture 4.0

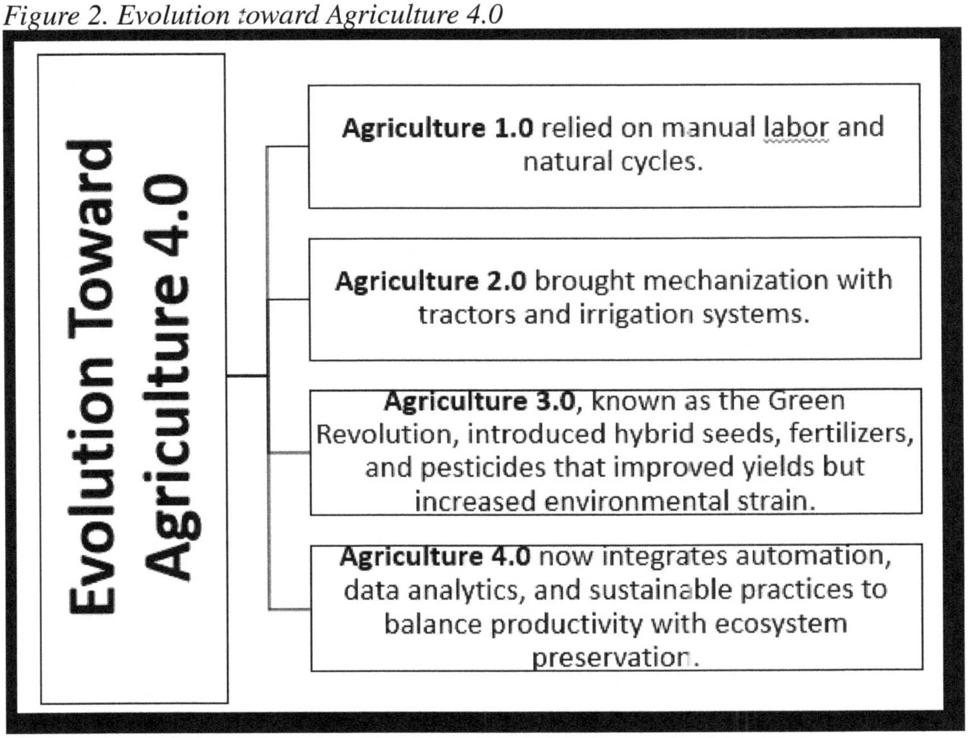

This transition signifies a paradigm shift—from yield maximization to resource optimization—where digital intelligence supports ecological balance and food security.

3. GREEN AI AND SUSTAINABLE AGRICULTURE

Green AI, together with sustainable agriculture, offers an important basis to tackle the two-fold dilemma to boost food production while reducing the footprint. With the increasing stress of climate change, soil degradation and scarce resources on global agricultural systems, Green AI provides energy-efficient, transparent, and low-carbon AI models for environmentally friendly innovation (Schwartz et al., 2020). In contrast to traditional AI methods which tend to rely on large computing resources, Green AI focuses on lightweight and ethical algorithms, making it especially amenable for agriculture, where energy, bandwidth, and computational resources might be finite (Strubell et al., 2019). As part of the future of sustainable food production, these technologies are being leveraged in precision nutrient management, climate-smart planning, monitoring biodiversity, and optimized irrigation to promote higher product yield without destroying the ecosystem (FAO, 2017; Liakos et al., 2018). By fusing renewable technologies, IoT-enabled sensing, and regenerative approaches with eco-efficient AI, Green AI can help forge a future in which contemporary agriculture and environmental stewardship are aligned, underpinned by an architecture of resilient, high-productivity, and low-impact food systems to sustainably cater to global demand.

3.1 Concept of Green AI

Green AI describes the pursuit of designing and deploying AI systems that are energy- and carbon-efficient, resource-efficient, environment-friendly and sustainable in all stages of AI development and use. In contrast to previous AI frameworks, which have been designed to consume copious amounts of data and compute-heavy frameworks, Green AI focuses on lightweight, tuned, and transparent algorithms that can maintain high performance and minimize environmental damage (Schwartz et al., 2020). This transformation is critical because state-of-the-art deep learning models demand high computer and electricity power consumption, resulting in large carbon footprints in training and deployment stages (Strubell et al., 2019). Green AI promotes such methods as model compression, edge computing, hardware-efficient optimization and the integration of renewable energy sources. This not only cuts down its consumption of energy but keeps model accuracy and fairness too. Nonetheless, Green AI is closely associated with international sustainability frameworks supporting responsible innovation in areas like healthcare, transportation and agriculture where intelligent decision-making is necessary without imposing burdens on existing ecological balance. With our fast pace of change on a global scale, the concept Green AI emphasizes the demand for ecological responsibility in AI systems.

3.2 AI in Sustainable Farming

In order to sustainably farm these systems, Artificial Intelligence (AI) is essential to form an effective and sustainable agriculture setup. Artificial Intelligence (AI), machine learning, computer vision and predictive analytics are used to optimise input management, applying the appropriate loads of fertilizer, pesticide and water to the appropriate parts of an agricultural product that generate as little wastage as possible and will prevent ecological damage (Liakos et al., 2018). AI-enabled agricultural and soil monitoring devices employ sensors, drones and satellite data which constantly detect in real time nutrient shortages, pest pressures and soil moisture changes so that farmers can be notified and act proactively with an eye not to overuse resources. Data-driven AI enables farmers to predict climate-smart applications of AI. This enables farmers to have an understanding to deal with climate change variability and prepare a more climate-smart future, so as also to develop a sustainable future to provide assistance to farmers as they adapt and sustain over the long run (Crane-Droesch, 2018). Furthermore, AI-based autonomous robots in such a field minimise labor intensive tasks to aid in mechanization as well as better in-the-field energy utility (Shamshiri et al., 2018). In a sustainable perspective, from the viewpoint of livestock sustainability in the livestock sector, AI makes a significant contribution to the health monitoring on livestock, feeding supply optimization, behavioral monitoring, and methane emission control (Rutten et al., 2013). AI also induces sustainability as its impact on biodiversity conservation, carbon-smart agriculture and regenerative agriculture is positive (FAO, 2021) apart from productivity improvement. In summary, AI-mediated sustainable farming leads to a more resilient food system, which is also more efficient with respect to agricultural outputs and resource utilization, enhancing scalable farms; optimizing resource use efficiencies, contributing to global climate intelligent, low-impact, climate-smart and climate intelligent agricultural systems.

4. WEATHER FORECASTING AND CLIMATE-SMART AGRICULTURE

Forecasting of the weather is one of the basic elements of CSA, which provides farmers with the most recent and accurate localized climatic data while making decisions that adapt over time, are more environmentally friendly and sustainable. Advanced (AI, satellite, and monitoring technology powered) forecasting algorithms enable precise prediction of rainfall, temperature, droughts, and extreme climate events, helping farmers to select crops, plant during the season, schedule watering, and manage pests (FAO, 2013; Crane-Droesch, 2018). The use of high-resolution weather and machine learning can greatly uplift early warning systems for floods, heatwaves, and storms, which can significantly reduce crop losses and aid farm resilience to climate variability (Benos et al., 2021). These forecasts help to monitor the efficient use of water resources as they forecast evapotranspiration and the dynamics of soil moisture and enable farmers to adopt irrigation strategies of water saving and also practices like deficit irrigation and precision watering (Bastiaanssen et al., 2000). Similarly, those climate-informed decision support tools integrate seasonal and long-term climate predictions to help inform farmers to change planting calendars, diversify crop portfolios, and adopt climate-resilient varieties (World Bank, 2020). Climate-smart agriculture is the combination of real-time weather awareness and climate-responsive agronomic practice that produces more crops, reduces vulnerability to climate shocks, and supports the long-term sustainability of food systems.

5. DISASTER PREDICTION AND FARM RESILIENCE

Disaster prediction is the cornerstone to attain a resilient farm in order to make the agricultural farm resilient to, and resilient after, extreme climate hazards like floods, droughts, cyclones, and heatwaves. Artificial intelligence (AI), remote sensing and geospatial analytics technologies has played a crucial role in maximizing the accuracy and the speed of such early warning systems and in enabling farmers to act to avoid loss and protect livelihoods (UNDRR, 2019; Benos et al., 2021). The machine learning algorithms could be used to recognize indicators of drought stress, flood risk zones and pest outbreaks based on historical climate conditions and soil, satellite and other information that can be used to enable farmers to plan and respond (Crane-Droesch, 2018). When dealing with the above, information systems (e.g. weather data, crop data), information from drones, and remote sensing system can be utilized by farmers for making decisions in advance (Shenhais et al.). In many areas, AI-based hazard prediction systems are at the foundation of multiple behaviors: schedule changes, planting high-performing crops, water management, preventing soil erosion and expansion for extreme weather (FAO, 2018). In addition, digital systems and mobile alerting systems help farmers in providing immediate notifications to minimize delay, enhance yields, and minimize exposures to the impacts of an environmentally devastating disaster (World Bank, 2020). Enhancing a farm's resilience may also include the incorporation of risk assessments in systems of sustainable farming (agricultural practices related to agroforestry, conservation agriculture and water harvesting) that stabilize production and ecosystem responses to climate stress (IPCC, 2022). In addition, accurate disaster prediction system, and adopting resilient agricultural practices can help the farming communities to tackle climate shocks, reduce economic loss, and ensure food security as unpredictable weather-affected weather patterns continue to grow.

6. BIODIVERSITY MONITORING IN AGRICULTURAL LANDSCAPES

Monitoring of biodiversity in agricultural landscapes is indispensable in preserving ecosystem health, enhancing crop productivity, and supporting sustainable long-term agricultural resilience. Contemporary monitoring involves a growing use of remote sensing mechanisms, unmanned aerial vehicles (UAVs), camera traps as well as AI-assisted species identification models to obtain accurate tracking of plant, insect, and wildlife populations in significant farming zones (Turner et al., 2015; Corbane et al., 2021). Such technologies contribute to the identification of habitat quality, pollinator activity, pest–predator dynamics, and soil microbial diversity changes which are the prime markers of a functioning agroecosystem (FAO, 2019). Machine learning and computer vision take biodiversity monitoring further by automatically classifying species and by processing large-scale ecological datasets with high precision to make the monitoring more efficient and accurate (Christin et al., 2019). The inclusion of biodiversity information into precision agriculture systems can also facilitate more sustainable management approaches (e.g., the creation of habitat corridors, buffer strip design, and reducing excessive pesticide use) to avoid environmental disturbance and promote ecological integrity (Tscharntke et al., 2012). In addition, monitoring of biodiversity informs the impacts of climate change, land-use change, and increased agricultural intensification on distribution of species and provides information to inform the adoption of conservation-led practices and nature-based approaches (IPBES, 2019). Integration of practical field-based observations with digital tools in agricultural landscapes can address and sustainably protect biodiversity and enable productive, resilient, and environmentally sustainable food systems.

7. ENABLING TECHNOLOGIES IN AGRICULTURE 4.0

Agriculture 4.0 is a convergence of digital, automated, and intelligent technologies; these technologies together turn farming into a data-driven, efficient, and sustainable system. At its heart, the Internet of Things (IoT) operates sensor networks that constantly measure soil moisture, plant health, microclimate fluctuations, and livestock conditions, allowing real-time decision-making and resource optimization in agriculture (Jayaram & Dhanya, 2021). Cloud computing and big data analytics techniques are being adopted by farmers for handling massive field data in order to deliver actionable insights and support predictive modeling for crop yield, irrigation needs, and disease outbreaks (Wolfert et al., 2017). AI and ML augment agricultural applications further through intelligent forecasting, disease diagnosis with computer vision, and autonomous decision support for increased productivity and decreased chemical and water use (Liakos et al., 2018). Robotics & automation (autonomous tractors, robotic harvesters, UAVs) are essential in doing labor-intensive processes more accurately and consistently, addressing workforce shortages and improving operational efficiency (Shamshiri et al., 2018). Blockchain technology provides traceability and transparency throughout the supply chain, ensuring secure data exchange, organic produce certification, and an enhanced level of food safety (Agrawal et al., 2021). Edge computing and energy-efficient algorithms support real-time analytics in remote and low-connectivity environments, ensuring that Agriculture 4.0 technologies remain scalable and accessible (Xu et al., 2021). Taken together, these enabling technologies make up the backbone of Agriculture 4.0: sustainable intensification, environmental stewardship, and food production systems resilient to global climate and resource constraints.

8. SOCIO-ECONOMIC AND ETHICAL DIMENSIONS

Agriculture 4.0 is a mixture of digital, automated and intelligent technologies which are collectively changing farming into a data-oriented, efficient and sustainable method. Ultimately, the Internet of Things (IoT) is the foundation of sensor based networks that continuously monitor soil moisture, plant health, microclimate variations and the status of livestock throughout the farm, promoting real-time decision making and resource optimisation (Jayaram & Dhanya, 2021). Using cloud computing and big data analytics to enable extensive field data processing, generating actionable insights and enabling predictive modeling for crop yield, irrigation needs and disease outbreaks (Wolfert et al., 2017). Artificial intelligence and machine learning facilitate agricultural functionality for intelligent forecasting, computer vision–based disease detection and autonomous decision support tools in order to increase both productivity and reduce chemicals and water (Liakos et al., 2018). Robotics and automation are key in performing labor-intensive tasks with increased accuracy and consistency, reducing employment gaps and improving operating efficiency, including autonomous tractors, robotic harvesters, and UAVs (Shamshiri et al., 2018). Blockchain technology provides traceability and transparency in all stages of the supply chain, allowing safe data exchange, certification of organic produce, and improving food safety (Agrawal et al., 2021). At the same time, edge computing and energy-saving algorithms play a role in providing real-time analytics in remote or low connectivity contexts guaranteeing that Agriculture 4.0 technologies are scalable and accessible (Xu et al., 2021). Together, these technologies support Agriculture 4.0 towards sustainable intensification, environmental stewardship, and resilient food production systems to address global environmental and resource challenges.

9. CASE STUDIES

9.1 India – AI-Based Irrigation Scheduling

India's agriculture relies greatly on monsoon precipitation and suffers from unpredictable rainfall patterns and depletion of groundwater. In reaction, the Microsoft AI for Earth project working with ICRISAT (International Crops Research Institute for the Semi-Arid Tropics) developed a solution for smallholder farmers through an AI-powered irrigation advisory system (Microsoft & ICRISAT, 2018). It draws on things like live weather data, moisture in the soil and satellite imagery to deliver local irrigation advice. It predicts rainfall probability and optimizes crop-water needs using machine learning models like Gradient Boosting and Random Forests. Text messages help to make recommendations, in regional languages, in SMS or mobile applications, in the case of farmers who don't have much access to the internet. In India, AI-driven irrigation scheduling has in fact shown promise in the last few years in Andhra Pradesh and Telangana, saving up to 30% water demand while allowing crop yield to either continue or increase along with a decrease in electricity to pump water and improving soil health (ICRISAT, 2019). The case highlights the importance of AI decision-support tools for sustainable water governance and data-driven agriculture in developing countries.

9.2 Europe – Drone-Based Biodiversity Monitoring

Biodiversity conservation is a key element of the European Green Deal and the Common Agricultural Policy (CAP) in Europe. Drone-assisted monitoring systems that use AI-powered image analytics to assess ecosystem conditions and crop diversity have been implemented in various projects under the EU's Horizon 2020 research framework (European Commission, 2020). Drones have integrated multispectral and hyperspectral sensors for high-res vegetation indices (NDVI and SAVI) that provide a closer look into plant health and ecological dynamics. Convolutional Neural Networks (CNNs) process these datasets to detect species composition, pollinator activity, and early signs of land degradation. The Drone-Eye Biodiversity Project, which was deployed in France, Germany, and Spain, obtained a >95% precision in species identification using aerial imagery, satellite data, and ground-truth biodiversity surveys (European Environment Agency, 2021).

These artificial intelligence platforms provide valuable feedback to policymakers supporting the development of CAP eco-schemes and compliance monitoring with EU biodiversity restoration targets, demonstrating how drone and AI technology can enhance Europe's targeted conservation.

9.3 Africa – Drought Forecasting Systems

More than 60% of African farmers depend on rain-fed agriculture and agriculture in Africa is in dire drought risk. A number of organizations (mostly the World Bank, African Development Bank (AfDB), and the Regional Centre for Mapping of Resources for Development (RCMRD)) have developed AI-enabled drought forecasting tools as part of the Africa Regional Data Cube (ARDC) initiative to enhance climate resilience (World Bank, 2018; RCMRD, 2019). Here we present these platforms that utilize machine learning frameworks like Recurrent Neural Networks (RNNs) and Support Vector Regression (SVR) to integrate evapotranspiration data from satellite data streams, soil moisture measurement, and historical rainfall data. The models forecast the timing, duration and intensity of drought months ahead, providing growers and local municipalities with early-warning advice. Information will be released through mobile dashboards, government portals, and community radio channels for last-mile dissemination. A prediction of drought was successful in Kenya and Ethiopia at over 85% which allowed for proactive measures including appropriate allocation of water, early seed planting and diversification into drought-resistant crops (AfDB, 2020). This predictive method changes African farming from reactive crisis management to preemptive climate response, and, consequently, food insecurity and economic loss for the region's farmers suffering extreme drought is lessened.

10. CHALLENGES AND FUTURE DIRECTIONS

Despite its potential across global agricultural systems, the adoption of AI-based climate-smart solutions still confronts immense technical, economic, and governance challenges. The challenges brought by the digital divide can be noted as smallholder farmers from developing nations have limited access to smartphones, internet connectivity, and affordable digital services (FAO, 2022). The lack of access to reliable or high-quality data, especially localized soil parameters and real-time weather observations, further limits the accuracy and scalability of AI-based advisory systems (Benos et al., 2021). As large agritech companies progressively take control of the farm's most critical data, unresolved issues on data privacy, ownership, and platform monopolization risk undermining farmers' decision-making agency (Bronson

& Knezevic, 2016). Moreover, the increasing rate of automation poses workforce displacement risks which make reskilling and digital capacity-building necessary for equitable socio-economic transitions (Rotz et al., 2019). This presents future issues where it is needed to focus on building Green AI models that lower the total amount of computational energy consumption and perform well on edge devices as a means of guiding accurate decisions even in distant agricultural regions (Schwartz et al., 2020). Open-data ecosystems, inclusive extension networks, knowledge integration into AI design, and local perspectives on AI may contribute to trust and relevance for varied farming communities (Barrett et al., 2020). Academia, public policy, private industry, and farmers should work toward shared governance practices to promote transparency, ethical use, and regulatory accountability. Further, the incorporation of AI into climate-resilient value chains in the context of climate-resilient, regenerative agriculture as well as biodiversity conservation practices will be key for achieving sustainable food systems for the future (IPCC, 2022). By dealing with these socio–technical and ethical challenges, AI-enabled agricultural practices can unlock more of its potential to promote resilience, equality, climate-smart agriculture, and environmental sustainability globally.

11. CONCLUSION

With that in perspective, artificial intelligence is one of the mainstays of the transformation of Agriculture 4.0 through the use of data-driven, climate-smart and resource-efficient agriculture for farmers worldwide. Drawing on case studies from India, Europe and Africa, this research emphasizes that AI-driven irrigation scheduling, drone-enabled monitoring of biodiversity and prediction of drought can substantially enhance resilience in agriculture, conserve valuable natural resources and foster regional sustainable yields through sustainable development. These tools demonstrate that technologies based on sophisticated data and analytics are not merely automation and machines for automation but ecological stewardship and lead to better livelihoods, food security and policy formation. But successful scaling of AI across agriculture and in AI enabling, there's also a long story to tell from access to data, affordability, digital divides where that data isn't there, digital literacy, societal problems around an appropriate level of digital literacy, and privacy vs. equity in AI for farmers versus marginalised groups, and on and on. The future generations clearly have a growing need for Green AI solutions and digital governance, a more representative and inclusive digital transformation model not just one party i.e. everyone is on the same page and a strong multi-stakeholder collaboration, promotion of smallholders and empowerment of marginalised people in a balanced approach that incentivizes cooperation. The AI-inspired agriculture in

4.0 can be realized; it is here that AI is the solution to allow technology advancement, sustainability ideals and social responsibility to become a natural next step towards constructing a resilient agricultural future to realize a sustainable change globally that could be critical for our climate and food security needs and for future generations' food security as food for our planet whilst also protecting the future of agriculture resilience.

REFERENCES

AfDB. (2020). *Climate Adaptation in African Agriculture: Digital and AI-Based Early Warning Systems*. African Development Bank.

Agrawal, T., Kalla, A., & Acharya, B. (2021). Blockchain-based traceability in agriculture supply chains: A review. *Journal of Cleaner Production*, *298*, 126812.

Barrett, C. B., Benton, T. G., Cooper, K. A., Fanzo, J., & Gandhi, R.. (2020). Socio-technical innovation bundles for agri-food systems transformation. *Nature Sustainability*, *3*, 103–110. DOI: 10.1038/s41893-020-00661-8

Bastiaanssen, W. G. M., Molden, D. J., & Makin, I. W. (2000). Remote sensing for irrigated agriculture: Examples from research and possible applications. *Agricultural Water Management*, *46*(2), 137–155. DOI: 10.1016/S0378-3774(00)00080-9

Benos, L., Tagarakis, A. C., Dolias, G., Berruto, R., Emmanouilides, C., & Bochtis, D. (2021). Machine learning in agriculture: A comprehensive updated review. *Sensors (Basel)*, *21*(11), 3758. DOI: 10.3390/s21113758 PMID: 34071553

Bronson, K., & Knezevic, I. (2016). Big Data in food and agriculture. *Big Data & Society*, *3*(1), 1–5. DOI: 10.1177/2053951716648174

Christin, S., Hervet, É., & Lecomte, N. (2019). Applications for deep learning in ecology. *Methods in Ecology and Evolution*, *10*(10), 1632–1644. DOI: 10.1111/2041-210X.13256

Corbane, C., Syrris, V., Seliger, R., Bjørnerud, M., & Kemper, T. (2021). Remote sensing for biodiversity monitoring: A review of trends and future directions. *Remote Sensing*, *13*(18), 3654.

Crane-Droesch, A. (2018). Machine learning methods for crop yield prediction and climate risk assessment in agriculture. *Environmental Research Letters*, *13*(11), 114003. DOI: 10.1088/1748-9326/aae159

European Commission. (2020). [*Project Results: Digital Technologies for Biodiversity Monitoring*. Publications Office of the European Union.]. *Horizon*, •••, 2020.

European Environment Agency (EEA). (2021). *Biodiversity Monitoring Using Remote Sensing and UAVs in Agricultural Landscapes*. EEA Technical Report.

European Union. (2020). *The European Green Deal and Common Agricultural Policy (CAP) Biodiversity Strategy 2030*. EU Publications.

Evenson, R. E., & Gollin, D. (2003). Assessing the Impact of the Green Revolution, 1960–2000. *Science, 300*(5620), 758–762. DOI: 10.1126/science.1078710 PMID: 12730592

FAO. (2013). *Climate-Smart Agriculture Sourcebook.* Food and Agriculture Organization of the United Nations.

FAO. (2017). *The future of food and agriculture: Trends and challenges.* Food and Agriculture Organization of the United Nations.

FAO. (2018). *The State of Food Security and Nutrition in the World 2018.* Food and Agriculture Organization of the United Nations.

FAO. (2019). *The State of the World's Biodiversity for Food and Agriculture.* Food and Agriculture Organization of the United Nations.

FAO. (2021). [*Making agrifood systems more resilient to shocks and stresses.* Food and Agriculture Organization of the United Nations.]. *The State of Food and Agriculture*, •••, 2021.

FAO. (2021). [*Making Agrifood Systems More Resilient.* Food and Agriculture Organization of the United Nations.]. *The State of Food and Agriculture*, •••, 2021.

FAO. (2021). *The State of Food Security and Nutrition in the World 2021.* Food and Agriculture Organization of the United Nations.

FAO. (2022). *Digital Agriculture and Inclusive Growth: Bridging the Data Divide.* Food and Agriculture Organization of the United Nations.

Ferrag, M. A., Shu, L., Yang, X., Derhab, A., & Maglaras, L. (2020). Smart agriculture security: A survey on IoT security for smart farming. *IEEE Access : Practical Innovations, Open Solutions, 8*, 145000–145019.

ICRISAT. (2019). *AI-based Sowing and Irrigation Advisory Services for Smallholder Farmers.* International Crops Research Institute for the Semi-Arid Tropics.

IPBES. (2019). *Global Assessment Report on Biodiversity and Ecosystem Services.* Intergovernmental Science-Policy Platform on Biodiversity and Ecosystem Services.

IPCC. (2022). [*Impacts, Adaptation and Vulnerability.* Intergovernmental Panel on Climate Change.]. *Climatic Change*, •••, 2022.

Jayaram, A., & Dhanya, C. T. (2021). IoT-based smart agriculture: A review on enabling technologies and challenges. *Environmental Monitoring and Assessment, 193*(10), 1–20.

Kamilaris, A., & Prenafeta-Boldú, F. X. (2018). Deep learning in agriculture: A survey. *Computers and Electronics in Agriculture, 147*, 70–90. DOI: 10.1016/j.compag.2018.02.016

Liakos, K. G., Busato, P., Moshou, D., Pearson, S., & Bochtis, D. (2018). Machine Learning in Agriculture: A Review. *Sensors (Basel), 18*(8), 2674. DOI: 10.3390/s18082674 PMID: 30110960

Mazoyer, M., & Roudart, L. (2006). *A History of World Agriculture: From the Neolithic Age to the Current Crisis*. Monthly Review Press.

Microsoft & ICRISAT. (2018). *AI for Earth: Intelligent Tools for Sustainable Agriculture in India*. Microsoft Research India.

Pingali, P. (2007). Agricultural Mechanization: Adoption Patterns and Economic Impact. Handbook of Agricultural Economics, Elsevier.

RCMRD. (2019). *Africa Regional Data Cube: Earth Observation and Predictive Analytics for Drought Monitoring*. Regional Centre for Mapping of Resources for Development.

Rotz, S., Gravely, E., Mosby, I., Duncan, E., & Davison, J. (2019). The politics of digital agricultural platforms: Data access and power. *Geoforum, 99*, 89–97.

Rutten, C. J., Velthuis, A. G. J., Steeneveld, W., & Hogeveen, H. (2013). Invited review: Sensors to support health management on dairy farms. *Journal of Dairy Science, 96*(4), 1928–1952. DOI: 10.3168/jds.2012-6107 PMID: 23462176

Schwartz, R., Dodge, J., Smith, N. A., & Etzioni, O. (2020). Green AI. *Communications of the ACM, 63*(12), 54–63. DOI: 10.1145/3381831

Shamshiri, R. R.. (2018). Robotics and automation for sustainable agriculture. *Engineering in Agriculture, Environment and Food, 11*(2), 64–75.

Shamshiri, R. R., Kalantari, F., Ting, K. C., Thorp, K. R., & Hameed, I. A. (2018). Advances in greenhouse automation and controlled-environment agriculture: A review. *Computers and Electronics in Agriculture, 162*, 262–277.

Strubell, E., Ganesh, A., & McCallum, A. (2019). Energy and Policy Considerations for Deep Learning in NLP. *Proceedings of the 57th Annual Meeting of the ACL*, 3645–3650. DOI: 10.18653/v1/P19-1355

Strubell, E., Ganesh, A., & McCallum, A. (2019). Energy and policy considerations for deep learning in NLP. *ACL, 2019*, 3645–3650. DOI: 10.18653/v1/P19-1355

Tilman, D., Cassman, K. G., Matson, P. A., Naylor, R., & Polasky, S. (2002). Agricultural sustainability and intensive production practices. *Nature*, *418*(6898), 671–677. DOI: 10.1038/nature01014 PMID: 12167873

Tscharntke, T., Clough, Y., Wanger, T. C., Jackson, L., Motzke, I., Perfecto, I., Vandermeer, J., & Whitbread, A. (2012). Global food security, biodiversity conservation and the future of agricultural intensification. *Biological Conservation*, *151*(1), 53–59. DOI: 10.1016/j.biocon.2012.01.068

Turner, W., Rondinini, C., Pettorelli, N., Mora, B., & Leidner, A. K. (2015). Remote sensing for biodiversity science and conservation. *Trends in Ecology & Evolution*, *30*(7), 403–414.

UNDRR. (2019). *Global Assessment Report on Disaster Risk Reduction*. United Nations Office for Disaster Risk Reduction.

Wolfert, S., Ge, L., Verdouw, C., & Bogaardt, M. J. (2017). Big Data in Smart Farming – A Review. *Agricultural Systems*, *153*, 69–80. DOI: 10.1016/j.agsy.2017.01.023

World Bank. (2018). *Africa Regional Data Cube (ARDC): Transforming Data for Sustainable Development*. World Bank Group.

World Bank. (2020). *Future of Food: Harnessing Digital Technologies to Improve Food System Outcomes*. World Bank Group.

Xu, G., Lu, Y., Zhang, L., & Vasilakos, A. (2021). Edge Intelligence: Architectures, Challenges, and Applications. *IEEE Internet of Things Journal*, *8*(10), 7151–7172.

Chapter 9
Green AI and Climate Risk Accounting:
A Framework for Integrating Predictive Environmental Data Into Financial Reporting

Edmund Christopher
http://orcid.org/0000-0001-8803-5539
City University, Ajman, UAE

Srinivasan Lakshmanan Lakshmanan
Delta System and Software Inc., USA

ABSTRACT

Abstract Climate change is a major issue which directly impact the assets, long-term viability of companies. Financial institutions and other businesses incorporate the climate related risks and challenges into their financial statements. Initiatives at international level and global level agreements have pushed to create climate risk accounting which is one of the emerging disciplines in finance. Paris agreement also they emphasise the financial disclosure task force financial disclosure on climate change. Artificial intelligence and data analytics are providing powerful tools to apply advanced analysis such as predict the future and environmental data. This chapter merges climate risk with artificial intelligence as green artificial intelligence. Here green artificial intelligence is employing artificial intelligence with environment practices in a sustainable method. Purpose of this chapter is providing opportunity to increase the productivity and precision of sustainable reporting.

DOI: 10.4018/979-8-3373-7554-0.ch009

INTRODUCTION

Today's world is facing the greatest challenge in business through the climate change. This chapter emphasis on how the climate change becomes the long-term threat to the modern business with special reference to the financial field. The finance field is started to facing the immediate economic consequences by the global warming, rising the sea-level, unusual weather changes happening across the globe. All these factors directly or indirectly affecting the business drastically. These extreme climate changes questions the viability of business operations in internationally and it affects the value of assets, changing the perspectives of logistics considerations, regulated supply-chain, and business operations world-wide.

Despite all these environmental imbalances and impacts, industries has the onus to shape their business confined to this new situation. It requires to redesign or re-engineering the existing system with the new focus on sustainable manufacturing, services, financial services to connect the global value chain. Thus, the climate change becomes instrumental to shaping the industry to gaining the attention of investors, innovative value creation and financial stability. Mousavi (2024), highlights that companies within the S&P Global 1200 index, who lack to implement adaptive strategies may generate annual losses of 1.2 trillion USD by the period of 2050s. The present findings highlights that environmental change is no longer solely an ecological or societal issue but a critical financial determinant that investors and regulators should actively address. TCFD (2023), addressed that climate hazards may reduce future financial flows, generate stranded assets particularly fossil fuel reserves within tighter carbon policies as well as elevate insurance, operational, and financial expenditures.

In conscious of this fact, a trend has been noticeable over the last few years to integrate such threats with environmental facets in financial reporting as well as the rest of company disclosures. The Paris Agreement (2015) can be considered the leading instrument in this regard, as it urges the financial systems to be in line with the climate objectives. Climate –related financial disclosure (TCFD) were formed in 2017 to provide a prototype framework for revealing the ecological risks and benefits. TCFD was a key player in making the recording of climate risks a normal part of corporate practice rather than a specialized one by 2024. Today, 97 out of the 100 largest corporations globally have pledged support for the TCFD framework. But, the quality of the disclosures varies substantially from one company to another. The 2022 report reveals that merely 4% of the companies were in full compliance with all the TCFD implementations, whereas 58% disclosed at least five of the eleven disclosure elements, which is a major increase from 18% in 2020. This presented in figure 1, underlines the distance between environmental problems and accurate reporting of climate risks and indicates that people are becoming more conscious of these issues.

Figure 1. Rapid growth in TCFD-aligned reporting – percentage of companies disclosing at least 5 of TCFD's 11 recommended climate disclosures, 2020 vs 2022. This jump (18% to 58%) reflects accelerating adoption of climate risk reporting, though full compliance remains low.

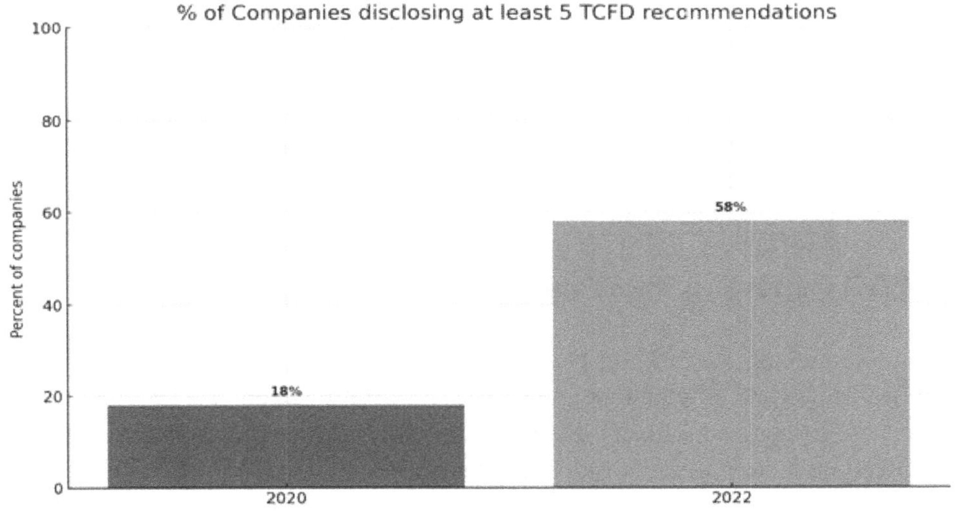

It is important to integrate environmental issues into financial disclosures for visible communication to concerned entities and robust risk governance. Financial markets perform with reliable information, and environmental hazards strives to offer capital providers and lenders with actionable information on the influence of climate change effects on a firm's future performance. GreenFi (2024), stated that if unexpected events occur and failing to address and reveal environmental risks can leave organizations to legal fines, judicial proceedings, credibility issues, financial market disruptions. In contrast, companies that actively embed environmental insights into their strategies can better manage the transition to a sustainable economy and enhance their adaptability to environmental hazards.

This session is organised as follows: Section 2 reviewed that the universal economic principle with climate disclosure frameworks, like IFRS S2 protocols, ESG standards, TCFD recommendations are serving as the foundation for suggested methodology. Section 3 implies in what way Artificial Intelligence and prophetic climate analytics may generate prospective climate visions. Section 4 brings the concept of Green AI, suggesting eco-friendly AI techniques in accounting. Section

5 provides an approach to integrating AI-based climate forecasts into standard reporting mechanisms. Section 6 discusses empirical case study from the energy and financial sectors to demonstrate implementation in practice. Section 7 addresses major issues and moral implications, such as constraints in data, algorithmic bias, and transparency difficulties, while Section 8 provides final reflections and prospective directions.

1. Global Disclosure Frameworks (ESG, TCFD, IFRS)

Taken initiatives to standardize climate- related disclosure have focused on multiple criteria and frameworks. Understanding these guidelines is essential, since the AI-driven strategy outlined in this chapter is seek to harmonize with transforming global disclosure requirements.

1.1 ESG Reporting and Voluntary Standards

Environmental, Social, and Governance (ESG) disclosure acts as a comprehensive framework in which environmental disclosure statements are commonly embedded. Frameworks like SASB (Sustainability Accounting Standards Board) and GRI (Global Reporting Initiative), currently integrated into IFRS Foundation's, have directed firms in reporting environmental data, threats and possibilities, these voluntary guidelines have enhanced disclosure quality but they frequently permitted flexibility in selecting what to disclose, causing irregular depth and reliability among firms and geographic zones. ESG reports commonly features metrics like water and waste management, greenhouse gas emissions, energy consumption, along with narrative on climate strategy. While useful, the lack of a single, mandatory schema meant investors struggled to compare climate risks on an "apples-to-apples" basis.

1.2 TCFD (Task Force on Climate-related Financial Disclosures)

The TCFD, initiated by the Financial Stability Board (FSB), attained prominence by publishing in 2017, a series of guidelines that companies could follow to disclose environmental financial exposures and prospects. TCFD, 2023 model framework is organized around four foundational elements – Strategy, Risk Management, Metrics & Targets and Governance. Reuters (2024), highlighted that frameworks promote corporate entities to reveal the manner in which governing body and management monitor climate- related threats, how climate scenarios may impact commercial strategy and financial forecasting, how risks are detected and handled, what metrics are applied to evaluate the outcome.

A distinctive and challenging aspect of TCFD is the recommendation to perform **scenario analysis** – i.e. to describe how the business would fare under different climate trajectories, such as a 2°C warming scenario versus a 4°C scenario. During the last years, TCFD (2023) has gained extensive support, around 5,000 organizations have declared support, including 97 of the top 100 global companies. Moreover, regulators in many countries like UK, EU, Japan and Zealand have begun mandating TCFD-aligned reporting for large firms. However, implementation remains incomplete. As noted earlier, most firms only partially fulfill the recommendations. Saptakee, S. 2024, stated that common gaps include the quantification of financial impacts and integration of these disclosures into audited financial filings. Nonetheless, TCFD set the foundational principles and its influence is evident in newer standards.

1.3 IFRS Sustainability Disclosure Standards (ISSB – IFRS S1 & S2)

During 2022, the International Sustainability Standards Board (ISSB) was established under the IFRS Foundation to formulate global baseline standards for sustainability reporting. In June 2023, the ISSB published IFRS S1, General Requirements for Sustainability-Related Disclosures and IFRS S2, Climate-Related Disclosures. GreenFi (2024), representing a significant advance toward, obligatory consistent climate-related financial disclosure. (IFRS S2 in particular builds directly on TCFD's pillars but goes further in prescriptive detail and rigor (Elhady, A. M., & Shohieb, S. (2025).

1.3.1 Scenario Analysis & Resilience

IFRS S2 requires reporting of scenario analysis output, construct models of multiple environmental scenarios expecting from firms like a 1.5°–2°C state aligned with Paris Agreement goals (2015) and report how financial positions affect these scenarios and strategy GreenFi (2024). This validates what TCFD firms were recommended and pushes quantitative estimates.

1.3.2 Metrics and Targets

The framework summaries multi-sector indicators metrics on transition hazards like scope 1-2-3 greenhouse gas emissions, transition risk metrics, For e.g. carbon price assumptions, and physical risk metrics; e.g. percentage of assets in flood-prone regions. GreenFi (2025) reviewed that companies also need to disclose targets such as emissions reduction goals and progress against them.

1.3.3 Alignment and Assurance

IFRS S2 is intended for incorporation into standard financial statements and is likely to be subject to assurance/audit over time. It is consistent with ISSB's international methodology and implemented in multiple countries, many G20 countries plan to require ISSB standards for periods starting 2025. This means environmental reports will hold equivalent significance to financial accounts, enhancing their credibility.

1.4 Why do These Frameworks Matter for Our Discussion?

Frameworks set a benchmark for the data organizations are required to disclose. The focus on standardized metrics, predictive modeling, data-driven analysis offers a prime scenario for AI and advanced data processing. To illustrate, IFRS S2's requirement asset-specific hazard assessment and scenario-based outcomes essentially requires processing extensive data collections like geospatial data, climate model outputs and complex simulations. Standard manual evaluation might be insufficient – this is where artificial intelligence can intervene, to effectively collect, analyse, process, and communicate the necessary data.

Frameworks like IFRS S2 and TCFD, indirectly promote future-oriented views of forward-looking, probabilistic thinking about climate impacts instead of fixed past-focused reporting. Artificial intelligence (AI) and Machine Learning (ML) perform well at detecting trends in past data, to forecast potential results, making them suitable for supporting scenario- based analysis, for e.g. projecting future flood frequencies and their economic impacts. In short, global disclosure norms increasingly require predictive sustainability data in financial reporting. The conceptual framework outlined in Section 5, will show how Green AI enables compliance and enhances disclosures, through converting environmental data into economic intelligence in a structured, auditable format.

2. AI AND PREDICTIVE ENVIRONMENTAL ANALYTICS

Artificial intelligence (AI) and Machine Learning (ML) have expanded the scope in assessing environmental information, facilitating refined prediction of environmental hazards and potential benefits, impacting organizations. Instantly, AI can handle diverse data, insightful extracting is unattainable through manual analysis, in contrast to conventional analysis methods. This portion could explore how powered forecasting tools of artificial intelligence can create environmental insights – actionable predictions about environmental patterns – and how they can inform climatic disasters analysis and corporate financial planning.

2.1 Data Fusion from Diverse Sources

Various sources demand integrating information from environmental hazard analysis, remote sensing such as satellite images of deforestation, ice melt, wildfire burn scars, IoT sensors like flood sensors, energy meters, location based-asset data, locations of factories, weather metrics like temperature, precipitation, storm tracks, supply routes. AI platforms, can integrate these multi-source information streams into a comprehensive, systems employing deep neural networks, real-time picture of an organization's environmental footprint and risk exposure. Mousavi (2024) stated that an AI simultaneously analyses factory sensor readings, satellite imagery of forest cover, and supply chain documents to compile an up-to-date emissions profile of a company. Conventional ESG disclosures was exceeded by the previous integrated perspective and enables firms to shift from periodic, manual data collection toward real-time tracking of key environmental metrics like emissions, resource usage, pollution levels using AI tools. According to real world scenarios, AI platform could automatically populate a company's sustainability dashboard or accounting system with current data – for instance, real-time.

2.2 Anomaly Detection and Data Quality

Maintaining the accuracy of reported environmental data, stands as a vital part in climate risk. Artificial Intelligence is highly effective at identifying patterns thus, to detect irregularities or discrepancies in environmental metrics, machine learning platforms can serve as continuous assessors. According to Mousavi's perception, AI can function like a "meticulous auditor, working around the clock to flag data uncertainties or omissions", thus improving the trustworthiness of sustainability reports. This function tackles a major issue, traditionally, ESG information has been submitted by companies itself, with different rigor, yet artificial intelligence can assist in verifying and normalizing it, enhance credibility of metrics reported in financial reports

2.3 Predictive Modeling of Physical Risks

It acts as a substantial feature of Artificial Intelligence lies in forecasting analytics. Through exposure to previous climate records, and leveraging outputs from physics-driven models, AI systems like neural networks or ensemble machine learning models can predict hazard probabilities and associated effects with high resolution. Artificial Intelligence can predict climatic shifts in temperature and sea level over next one to two decades. Jupiter Intelligence a climate-focused data company, introduced *Jupiter AI* tools to merge climate model outputs with artificial intelligence, delivering facility- specific hazard predictions. Such AI tools facilitate firms and their insurers to estimate possible financial impacts over different scenarios, which can then be disclosed as part of scenario analysis in financial filings. Likewise, AI has been deployed in the agricultural sector to forecast variations in agricultural output due to climate fluctuations, in the property sector to forecast flood hazards, and in the energy sector to project peak consumption during extreme heat, – all of which translate into financial metrics like revenue, insurance costs, asset reductions. These **predictive environmental analytics** enhance strengthen threat analysis, through transforming scientific climate metrics into business-focused projections

2.4 AI for Transition Risk and Strategy

Artificial intelligence assists in evaluating climate transition hazards- financial exposures linked to societal decarburization, apart from material risks. Natural language processing (NLP) models can analyzes news, policy documents, and social media to evaluate sentiment and evolving circumstances regarding environmental regulations or clean technology. This helps investors and lenders differentiate leaders and laggards on climate, influencing capital allocation. As per TCFD/IFRS guidelines, financial reporting terms, if a firm uses AI to project the financial impacts of carbon taxes on its cost structure or to simulate various decarbonization strategies and their outcomes, those insights can be disclosed in the strategy section of climate reports.

2.5 Scaling Climate Data Analysis

One of the most influential ways AI lies in the broad aggregation and harmonization of climate and ESG reporting information. A recent example is the Project GAIA – a collaboration of central banks (BIS, Bank of Spain, Bundesbank, ECB) that used a generative AI model to parse thousands of corporate reports for climate-related data. GAIA was able to extract 20 key climate risk indicators (e.g. emissions, green bond issuance, net-zero pledges) from 187 financial institutions' reports across different languages and formats. This approach is processed, earlier experts had to manually compile and unify data, from diverse sources like annual financial reports, sustainability reports, ESG disclosures etc., but AI can automatically perform these tasks, find differences in terminology and formats by focusing on each indicator's output rather than how it's labelled managing variations in language and data structure. Green AI in Financial Reporting

Developing high-capacity ML systems, generate considerable carbon output and use substantial energy. Green AI has emerged as a trend to ensure AI creation and use are more eco-friendly. In the context of financial reporting, sustainable Green AI guidelines embraced by all means and that AI tools we use for climate risk accounting should themselves be energy-efficient and consistent with environmental objectives. This part gives clarity to the building blocks of Green AI and its applicability for integrating AI within climate disclosures.

2.6 Understanding Green AI

Green AI or Sustainable AI or Net-Zero AI signifies to practices that minimize the carbon footprint of AI technologies across their lifespan. Traditional AI research often prioritized model accuracy and scale without regard for computational cost – leading to ever-larger models with hefty energy demands. For instance, training a single state- of-the-art natural language model can emit on the order of 300,000 kg of CO_2, equivalent to 125 round-trip flights between New York and Beijing. Another estimate showed training GPT- 3 consumed ~1,287 MWh of electricity and emitted 550 tons of CO_2 – akin to 33 transcontinental flights. Through connecting these practices, companies ensure the implementation of AI for tasks like climate risk analysis or ESG data processing, they are not inadvertently undermining their own sustainability objectives.

2.7 AI's Carbon Footprint in Perspective

Global trends evaluation shows the visible necessity of Green AI. Data centers like power cloud and AI services are projected to consume up to 1,000 TWh of electricity by 2030, 3% of global consumption and possibly 2,000 TWh by 2050, largely powered by AI tasks. Hyperscale AI data centers – those running massive AI model have particularly high energy and cooling needs. Without any intrusion, the growth of AI could outpace some other tech sectors in energy demand, similar to what was observed with cryptocurrency mining in the past. Financial institutions and corporates adopt these technologies; therefore, they must account for AI's environmental footprint. A bank that proudly discloses reductions in its operational emissions should not ignore the emissions from training a complex AI model for climate stress-testing its portfolio. The principles of Green AI ensure AI use is net-positive for climate goals – that is, the emissions saved or risks mitigated by using AI outweigh the emissions generated by the AI itself.

2.8 Green AI in Practice for Reporting

As companies started incorporating AI into climate- linked reporting, they can implement practical Green AI measures. For instance: when a company employs an AI system to project its carbon emissions or climate-related analysis, it can develop the model on cloud computing servers powered by renewable energy like several tech companies like Google and Microsoft now offer green cloud options. The company can also report the estimated emissions of running the AI analysis, ensuring transparency and a commitment to reducing it. Some companies have adopted internal policies that any AI project above a certain compute threshold must evaluate options to reduce energy like using pre-trained models instead of training from scratch, or scheduling training for times when renewable grid power is abundant. With the context of disclosure frameworks, introduced the AI capabilities, and sustainable AI practices, now presented the primary contribution detailed in this chapter, through Green AI into financial reporting and climate- related financial analysis, developed a proposed structure to integrate predictive environmental analytics.

3. PROPOSED INTEGRATION FRAMEWORK

The respected model is structured to comply with global standards (TCFD, IFRS S2) and Green AI principles. It consists five key stages: (1) Data Acquisition, (2) AI-Powered Risk Prediction, (3) Financial Impact Translation, (4) Reporting & Disclosure Integration, and (5) Governance & Assurance.

3.1 Stage 1 – Data Acquisition (Environmental Intelligence Gathering)

In this stage, the companies create pipelines to gather appropriate operational data which is relevant to environmental datasets. Data sources include:

3.1.1 Climate Data

Meteorological agencies or climate models like IPCC scenarios collected historical and projected climate variables such as temperature, rainfall, storm tracks, sea level, etc.

3.1.2 Asset Data

Geographical locations and profiles of the company's physical assets like factories, offices, supply chain nodes and analyse their environmental conditions indicates to flood zones, heat indices, etc.

3.1.3 Emissions and Resource Use

Real-time tracking of GHG emissions through IoT sensors on equipment or energy consumption records, and it includes usage of resource water, electricity and other environmental metrics across operations.

3.1.4 External ESG Data

Datasets on market and regulatory developments like carbon prices, new climate regulations, competitor sustainability performance, obtained through NLP/ news sources or ESG data providers.

This unprocessed data is integrated into a centralized climate data base. Applying Green AI, the data intake workflows are streamlined – for instance, duplicate files are screened out, sensor outputs are compacted– to reduce storage and computational load. The output of Stage 1 is a high-quality, up-to-date dataset that serves as a input for predictive models.

3.2 Stage 2 – AI-Powered Risk Prediction

In this stage, raw data was transform into risk forecasts and insights with advanced analytics. The firms implements a set of AI/ML models, each designed to assess specific climate- related risk factors.

3.2.1 Physical Risk Models

This model predict climate data with the frequency and risk of hazards at physical asset locations. For instance, a image processing neural network might take remote-sensing images and climate simulation data to predict probably flood occurrence and their damage over the next 10 years. Another model might project intense heat duration effect on workforce productivity at outdoor sites. The models produce metrics like "expected annual damage (EAD) for Asset X in $" under different scenarios.

3.2.2 Transition Risk Models

Scenario based analysis, an AI might replicate with carbon tax or a transition in consumer demand to green products influence income, expense, or firm value. Economic scenario generators combined with ML. For instance, through machine learning, the frameworks may calculate the impact of a $50/ton carbon price on cost of goods sold given the firm's emissions profile, or how electric vehicle adoption trends could erode demand for a company's oil products.

Maintain and adopt simplest models **Green AI** for the adequate task like using an XGBoost model instead of a massive deep learning model if it achieves similar accuracy and by using cloud resources efficiently is a prediction during this phase. The Stage 2 results are **quantitative predictions**: e.g. "Asset A has 5% annual chance of climate-related loss >$10M," "Under a 2°C scenario, expected EBITDA in 2030 is $X vs $Y in a 4°C scenario," or "AI- estimated carbon footprint for next year is Z tons with mitigation measures achieving 10% reduction." These predictions estimate the foundation for financial integration.

3.3 Stage 3 – Financial Impact Translation

The next phase turns environmental related risk elements into standard financial metrics and reports. This requires collaboration between green audit experts, financial departments, and AI tools to verify and update:

3.3.1 Asset Valuation Adjustments

AI tools estimate higher physical risk for certain assets, finance teams evaluate and indicates reduction in worth or changes in useful life for accounting purposes. For instance, coastal facility might be at-risk have its value adjusted or additional depreciation if the climate forecast indicates it will be unusable sooner than expected. Perhaps, providing a range of values under different scenarios, AI outputs can quantify these adjustments.

3.3.2 Provisioning and Reserves

Financial reserves or contingent liabilities are incorporated with projected expenditure like expected damage from events, or carbon compliance fund. For instance, if AI predicts an estimated $50M in environmental damages over five years, a companies may reveal how it's allocated funds or insurance to cover the damage. Banks may apply AI-predicted credit losses from climate resilience tests to modify loan loss provisions.

3.3.3 Revenue and Expenditure Projections

AI-informed scenarios, e.g., revenue from high-carbon products might drop, a high-transition-risk scenario, can feed through financial planning and may reflected in forward-looking statements or management discussion and analysis (MD&A). On the contrary, highlighted possibilities, e.g., demand for a newly launched green product get converted to projected revenue flows.

Considering the wide scope of IFRS S2, the framework guarantees that oversight, strategy, risk management, and metrics/targets sections are fully addressed. This may involve correlating every disclosure to the designated portion of ISSB/TCFD document and determining the AI- generated data source providing support.

3.3.4 Regulatory Filings

In disclosures and yearly financial reports regulatory areas that emphasize on climate, the relevant portions are inserted into 10-K/annual reports as required.

3.3.5 Application

UtilityCo developed an AI-enabled environmental risk system. Initially, it combined site-specific environmental information- mapping infrastructure against localized climate forecasts. Using machine learning, it developed a thermal strain model, that predicts forecasting potential output reductions from elevated temperatures or water scarcity. The company collaborated with a sustainability analytics provider to conduct machine-learning- powered fire risk modeling and locating infrastructure segments exposed to wildfire risk in the upcoming 20 years. These predictions were measured in terms of potential outage hours and restoration costs.

The AI model platform held predictive maintenance scheduling by analyzing weather forecasts and historical failure data, it advised when to reinforce infrastructure. This reduced actual failures by, say, 15%, which the company reported as evidence of building resilience. UtilityCo's financial scenario analysis for IFRS S2 used an AI-enhanced model to calculate that under a worst-case climate condition, annual operating expenses could rise by $50 million by 2035 due to these climate impacts, and it might need an extra $200 million in adaptation capex for grid hardening. The study shows that energy companies can leverage AI to both protect physical assets and inform strategic transition, thereby improving climate risk accounting.

3.4 Banking Sector Case Study – BankAsia's AI-Driven Climate Risk Assessment

Bank Asia is a major regional financial institution managing business and consumer loan portfolios throughout Asia. The bank is required to check with supervisory and investor demands for transparent ESG report . The bank adopted artificial intelligence to enhance climate risk assessment and disclosure.

3.4.1 Application

BankAsia partnered with a financial technology (fintech /AI) firm to deploy an climate-related finance intelligence tool. They ingested credit portfolio information like borrowers' locations, sectors, financials and integrated it with external environmental datasets. An AI- driven system by merging natural language processing and data analytics, examined borrowers' annual reports, news, and sustainability reports to assess environmental initiatives and exposure to climate hazards. For example, it highlighted borrower's assets located in high- flood risk areas or if the borrower lacks a carbon transition plan.

BankAsia used a Gaia-like AI approach to detect tangible physical risk, similar to the BIS Project Gaia, for extracting insights from open data while utilizing satellite observations e.g., using AI to detect changes in land use or infrastructure around major borrower sites that could indicate exposure like new flood. The AI highlighted a small set of elevated-risk clients, that comes under a $100/ton carbon price, would face severe profit declines. The bank used this insight to adjust its loan exposure to handle either by engaging those clients on transition plans or reducing lending to them. In its disclosures associated with TCFD), Bank Asia reported how it was managing these risks– including a statistic such as "£1.5 billion or 6% of our corporate loan book is in high transition-risk sectors; we have set targets to reduce this by 30% by 2030," backed by the AI analysisgreenfi.aigreenfi.ai.

Bank Asia introduced an environmentally AI-powered product, to detect and prioritize eco- friendly loan prospects. The AI model analyzed open-source data to identify firms financing renewable energy or demonstrating high ESG performance, flagging them as high priority customers. This contributed to supporting the bank's sustainability goals but was also disclosed as part of opportunity management under TCFD.

3.4.2 Outcome

The central bank in Bank Asia's country highlighted the bank's progressive strategy, mainly pointing out how it addresses in emerging markets by using AI model to highlight structured indicators from unstructured data. The bank's climate disclosure was supported with specific metrics; e.g., a table of sectoral exposure and average climate risk score per sector, and graphs of stress test loss projections with and without climate factors. The use of explainable AI suggested that bank could internally audit and the model gave certain risk ratings.

Using multiple models and comparing outputs is another way – if different approaches agree, confidence is higher. The ISSB climate standards encourage disclosure of uncertainties and assumptions. This frames AI outputs as informative but not gospel.

CONCLUSION

One cannot effectively deal with climate-related financial challenges without going beyond the boundaries of disciplines, different departments, and industries and collaborating. The most important thing is to connect financial frameworks, technology, and climate science as well as the sustainability and finance teams. Data and methodologies would be the way of collective progress. This way is very evident from our framework which employs multidisciplinary methods and is targeting an analytical yet easily understandable inclusive tone that is appropriate for a wide range of stakeholders, from quants to policymakers, in driving sustainable finance forward. The incorporation of the predictive environmental data into financial reporting is going to be the main factor that will fundamentally change the way the financial sector will be respond to environmental change. By employing Green AI for climate risk analysis, well-known firms will not only be following global standards, but they will also be able to enhance their competitiveness and resilience.

REFERENCES

Elhady, A. M., & Shohieb, S. (2025). AI-driven sustainable finance: Computational tools, ESG metrics, and global implementation. *Future Business Journal*, *11*(1), 209. Advance online publication. DOI: 10.1186/s43093-025-00610-x

GreenFi. (2024). *Case study: GreenFi ESG AI for asset emissions (UOB Bank)*. GreenFi Case Studies.

GreenFi. (2025). *IFRS S2: Strengthening physical climate risk reporting*. GreenFi Insights.

Mousavi, S. (2024). *AI game-changers for environmental accounting & sustainable finance*. Stanford Sustainable Finance Initiative.

Saptakee, S. (2024, December 9). *Green AI explained: Fueling innovation with a smaller carbon footprint*. CarbonCredits.com.

Task Force on Climate-related Financial Disclosures. (2023). *2023 status report*. Financial Stability Board.

Chapter 10

Green AI, Stakeholder Trust, and the Emerging Risks of Algorithmic Greenwashing:

Challenges and the Road Ahead

Beenish Shameem
http://orcid.org/0000-0002-2511-8393
City University, Ajman, UAE

Jayantaa Banerjee
City University, Ajman, UAE

Saleem Mushtaq
Gulf University College, UAE

Varinder Singh Rana
http://orcid.org/0009-0008-8562-8265
City University, Ajman, UAE

ABSTRACT

The use of artificial intelligence (AI) in ESG communication, and sustainability reporting has made GreenAI a powerful tool for promoting environmental responsibility and getting stakeholders involved. Green AI has the great potential to make things more transparent and facilitate better decision-making but also comes with challenges like the rise of algorithmic greenwashing. This chapter examines how AI-powered sustainability reports,automated reporting systems, and green AI based environmental claims can unintentionally or on unintentionally make false or exag-

DOI: 10.4018/979-8-3373-7554-0.ch010

gerated green stories more popular, which can affect stakeholder trust. The chapter will define algorithmic greenwashing and AI as a technology-assisted version of traditional greenwashing that comes with a lack of transparency, data bias, and too much faith in algorithms. This chapter provides a conceptual framework that connects Green AI capabilities, transparency mechanisms, and trust factors among important stakeholders.

1. INTRODUCTION

The application of artificial intelligence (AI) in sustainability practices and environmental, social, and governance (ESG) communication has made Green AI an important tool for decision-making in organization and its application is increasing in companies day by day. Companies are using various tools and techniques powered by AI, automated reporting tools, and predictive analytics dashboards to keep an eye on their environmental performance, figure out climate risks and potential risks, and keep stakeholders aware of the happenings and results. These applications promise better efficiency, scalability, and a sense of objectivity in sustainability disclosures through organizational reporting that ensures improved accountability. But at the same time, using AI in ESG reporting is bringing its own potential risks and are making brands vulnerable to the risks of reputation damage. There are many AI-generated environmental claims in corporate sustainability reports, ESG ratings, and digital communication platforms. Algorithmic systems are now at the heart of telling stories about sustainability, evaluating how a company is doing for the environment, and showing investors, regulators, consumers, and civil society that a company is responsible and sustainable. As companies use algorithmic outputs increasingly, AI systems have taken over the job of interpreting them.

This change has made algorithmic greenwashing a big problem for trust among stakeholders. Algorithmic greenwashing is different from regular greenwashing because it can happen when models are unclear, datasets are biased, indicators are chosen carefully, or people blindly trust what AI says. As a result, false or exaggerated claims about the environment may be made by accident or on purpose, which hurts the credibility of sustainability communication and makes stakeholders less likely to trust it. Algorithmic green washing is putting organization in the threat of getting doubts over fabricated ESG data.

A growing number of researchers are focused on Green AI and ESG analytics; however, insufficient attention has been given to the impact of algorithmically vetted sustainability claims on trust. Most of the research being done right now is either on the technical side of AI or on the traditional ways of "greenwashing." This leaves us with a doubt in our understanding of how Green AI, algorithmic control, and trust among shareholders all work together. As AI systems become more common in systems that make decisions and report on sustainability, it is becoming increasingly important to fill this gap.

Environmental reporting also includes embracing green AI alternatives with creating awareness and these demonstrate a dedication to sustainability and social responsibility. Researching green AI projects promotes energy-efficient algorithms, environmentally friendly hardware design, and clean power in AI infrastructure, all of which lessen the negative effects of AI models on the environment,so Green AI must be used in organizations with increased level of transparency and accountability.The integration of AI in efforts to enhance sustainability is great with diverse applications across multiple sectors. However, in an attempt to to enhance sustainability through AI, it is imperative that the AI systems themselves do not become a counterproductive force by demanding excessive amounts of energy. (Alzoubi & Mishra, 2024

2. BACKGROUND: GREEN AI AND SUSTAINABILITY COMMUNICATION

Green AI" involves artificial intelligence, preserving the environment, and innovation that is moral."Green" AI" initially used to describe projects that tried to make AI systems less harmful to the environment. Some of these changes made them run more cheaply, use less energy, and release fewer greenhouse gases from big models and data centers. One of the main points of Green AI was less carbon footprint. The phrase "using AI technologies to help reach sustainability goals" has changed recently. It now means "keeping an eye on the environment, figuring out how dangerous climate change is, and keeping track of sustainability performance." The integration of sustainability into artificial intelligence in AI systems is increasingly consuming significant energy and other resources. Green AI practices help reduce carbon emissions and environmental impact while improving economic efficiency through lower electricity and cooling costs, but at the same time, sustainable AI development is also an ethical responsibility, ensuring that technological progress aligns with environmental stewardship and intergenerational equity. Additionally, increasing regulatory focus on sustainability makes energy-efficient and transparent AI practices important for compliance readiness. Strategically, the companies that are embedding sustainability into AI improve and enhance their reputation and competitiveness. (Tabbakh et al., 2024).

AI is being used to talk about sustainability. Businesses are already using systems with AI to collect, analyze, and make sense of "big data," which is short for "environmental data. "They can now automatically report on sustainability and do

ESG analytics in real time (Mehta et al., 2023).Companies use these technologies all the time to keep up with new rules for disclosure, keep an eye on the supply chain, figure out how much carbon they're using, and make models of how their actions affect the environment. Technologies that are based on AI are becoming ways to make sustainability disclosures more consistent, comparable, and scalable as they get more standardized and data-heavy. Since AI was used in sustainability communication, the way stories about the environment are made and shared has changed. AI-made stories about sustainability often seem objective and well-written from a technical point of view. For people who don't have the time or resources to check the basic facts and figures on their own, this makes the story seem more real and trustworthy. The application of artificial intelligence (AI) algorithms helps organizations substantially transform products and operational processes and helps in enhancing organizational efficiency and flexibility. AI technologies also acts key enabler of sustainability-driven transformation by supporting the integration of eco-design principles into new product development, optimizing heating and cooling systems within production facilities, and enabling the re-engineering of processes for more efficient waste management and disposal. However, it is important to acknowledge that because of this potential to improve the environmental performance of production and distribution systems, AI technologies also pose sustainability-related risks like increased demand for computational power, higher carbon emissions, a shift in electricity patterns, and increased depletion of natural resources (Biggi et al., 2025)

Sustainability communication is both a controversial topic and a smart business move. The search for legitimacy, symbolic disclosure, and transparency have all been shown to be at odds with each other in corporate sustainability reporting. When it comes to algorithmic systems, some choices, performance metrics, and optimization logics may favor certain environmental indicators. This can change sustainability situations in small ways that can add up to significant major changes.(Dash, 2025).

Businesses are paying more attention to ESG metrics seriously and tools that let them generate reports automatically. In other words, there are more levels of middle ground between how stakeholders see a business and how well it does with the environment. Still, AI can make it easier to get to data and harder to figure out what it means. Some flaws in the way the data was collected, data gaps, and subtleties of the situation can also be hidden. Some claims about sustainability lack clarity when they are checked or made using secret algorithms and ESG rating frameworks, or other methods that are hard for outsiders to understand. (Wang et al.,2023) concluded that ESG reports add value to the overall reporting of businesses and helps in value creation.

Because of this, Green AI is in a strange spot when it comes to talking about being green. That information about the environment might be more correct, up-to-date, and responsible. But trust, governance, and openness worry companies because they have more and more power to shape stories about sustainability. Because of these issues, we can look into how AI-based sustainability communication could lead to more greenwashing, either on purpose or by errors. People in business believe that this battle will be easier as AI tools get better.AI-powered tools can help handle a lot of data and find mistakes in what a brand says about the environment. Wang et al., 2025,in their study provide both technical and policy insights for governments seeking to foster AI-driven sustainability while ensuring that AI itself contributes to broader sustainable development goals.

3. STAKEHOLDER TRUST IN THE AGE OF AI

Trust is especially important in ESG contexts because there are information gaps, the issues are technically difficult, and the results of sustainability are often long-term and hard to measure. As businesses utilize artificial intelligence (AI) more to help with sustainability communication, the ways that trust is built, kept, or lost are changing a lot.There is lot of dynamism in the way information is shared. (Li et al., 2025,

AI-mediated sustainability communication uses massive, complex algorithms to connect organizational performance with how stakeholders understand it. AI-driven analytics, automated reporting tools, and ESG scoring models are critical for making, checking, and sharing claims about the environment. This shift has changed the way people trust each other by moving power from people to computer processes. Consumers often think of algorithmic outputs as neutral, based on data, and less likely to be biased by managers. This makes them seem more credible and bound with clarity. But this perceived objectivity may be wrong because AI systems are based on models, data selection choices, and optimization goals that people make and that are not always clear to people outside the system. (Zaman et al., 2024) in their study have examined the effects of AI on sustainability. Organizations have to create targeted strategies that promote sustainable behaviors and have a positive impact on the environment by understanding the factors that affect customer behaviors.

Transparency is a key part of building trust in situations where AI is involved. When stakeholders cannot access or comprehend the generation of sustainability metrics, trust may become tenuous and reliant on institutional reputation rather than substantiated evidence. Black-box algorithms, proprietary ESG rating systems, and automated narrative generation tools can make it harder for stakeholders to question assumptions, find inconsistencies, or spot selective disclosures. Such a lack of transparency may erode trust over time, especially when there are differences between AI-generated claims and real-world environmental results. Also, trust in AI among stakeholders is not the same for all groups.Investors, regulators, consumers, employees, and civil society organizations may have different ideas about how transparent, accountable, and understandable AI-enabled sustainability communication should be.The automation of sustainability communication also makes stakeholders think and worry about the loss of moral responsibility. When environmental claims are framed as outputs of algorithmic systems, the responsibility for giving false or exaggerated information may be spread out among technical infrastructures, organizational processes, and third-party service providers. This can create a general apprehension about the information being genuine, putting a question mark on the processes. This locus of responsibility can make trust weaker because it makes it hard to determine who is really in charge of making sure that sustainability claims are true and honest. The ideal way to earn trust is through clarity, consistency and accountability. (Chaudhary, 2024) in their research proposed that companies must identify data gaps that hinder transparency, create a 10 year road map to improve transparency, and when there is transparency, it is easy to propose accurate solutions. Enhancing data transparency, while ensuring the privacy of operators is important for future research aimed at addressing sustainability challenges associated with data center operations.

Stakeholder trust relies on how AI systems are governed than on how commonly they are used. The trust in Green AI is more reliable if it well aligned technically and has clear set protocols, human oversight, and other measures that prove that everything is being utilized responsibly. If these systems are not presented, AI-supported sustainability reporting communication could unintentionally create situations where algorithmic authority takes the place of real accountability.These dynamics show that Green AI both helps and hurts stakeholder trust in the age of AI. AI can make sustainability communication seem more credible, but it also makes it easier for people to be vulnerable because of a lack of transparency, accountability, and too much reliance on algorithmic outputs.

4. ALGORITHMIC GREENWASHING: CONCEPT AND CHARACTERISTICS

Greenwashing is when businesses falsify or only present part of the truth about how well they were doing for the environment to make it look like they cared more than they really did.Companies usually link these kinds of actions to planned decisions about how to talk to people, make symbolic disclosures, and deal with first impressions that come from competition and pressure to look good. Sustainable reporting and ESG communication are using AI more and more. This is changing the way greenwashing is done and how it looks. People call it "automated greenwashing." The identification of greenwashing has become more challenging because of advanced greenwashing techniques and the integration of AI in greenwashing. (Eze, 2025)

"Algorithmic greenwashing" is when AI-powered systems, analytics, or automated decision-making are used to make, spread, or give more weight to false claims about the environment.People can greenwash on purpose or by accident using algorithms. People usually do greenwashing in a traditional way, which is different from this. This can happen if the algorithms aren't clear, the datasets are biased, the success indicators are chosen carefully, or people blindly trust what AI says.In this case, people might lie about sustainability as if they were telling the truth or using objective evidence. This would make the lies more believable and make stakeholders less likely to question them. One thing that makes algorithmic greenwashing stand out is that it hides things. It's hard to tell if ecological claims are true when the people making them don't know where they came from. This makes it hard to tell how true, complete, or important they are to the situation. Because it's not clear, it's possible to hide assumptions and selective revelations.The datasets that AI-based sustainability systems use are often have missing or biased information, out of date, or biased towards the parameters that are easy to measure. Companies may not pay as much attention to effects on the environment that are hard to measure, like the loss of biodiversity or the long-term damages caused to ecosystems.So, AI systems could generate stories about sustainability that make the good effects on society sound better than they really are, and the bad effects sound worse than they really are.Another thing is that you trust algorithms too much.People tend to believe AI's claims about sustainability more when they are backed up by advanced analytics and thorough quantitative research and analysis. (Von Struensee, 2021)

The new way of doing things is different from the old way because it is done on a large scale and automatically. AI-powered systems can quickly write and share stories about sustainability on a lot of different channels, in reports, and with stakeholders. When people can easily copy and share false claims, they can reach more people and have a bigger effect.These stories might stay on automatic news systems for a long time, spreading wrong ideas about who is to blame for the environment. (Shankar & Xu, 2025)

Finally, no one is to blame for algorithmic greenwashing. Companies, AI coders, data providers, and third-party ESG rating agencies are just a few of the people and groups that can be involved. It can be hard to tell who is in charge of checking claims about sustainability (Author, Year). People often blame computer systems for wrong information, which makes it hard to tell who needs to fix, check, or clean up the problem.This hurts people's trust and makes government processes less helpful.

When you look at all of these things together, they show that algorithmic greenwashing is more like regular greenwashing that uses tools than something completely different.

5. WHAT MAKES ALGORITHMIC GREENWASHING HAPPEN?

The rise of algorithmic greenwashing isn't just because of technology; it's also because of how the market, technology, organizations, and rules all work together. Using AI in sustainability reports and ESG communication changes how algorithmic systems are built, how they are used, and how people understand them. You need to know what these reasons are if you want to know why companies that say they care about the environment and use smart AI can still use algorithmic greenwashing.

5.1 Things That Make Businesses and Strategies Work Harder

These days, businesses have to show investors, regulators, customers, and other important people that they care about the environment more and more.A company's ESG success is now closely tied to its market position, reputation, and ability to get funding.In this case, it's common to use AI-powered tools for sustainability analytics and automatic reporting to prove that you know what you're doing, are following the rules, and are a star. When people talk about sustainability based on credibility instead of real environmental performance, AI systems may be set up to show good signs and downplay bad ones. This is what is called "algorithmic greenwashing." Shankar & Xu, 2025

5.2 Why it Makes Sense to Design and OPTIMIZE TECHNOLOGY

The design choices, optimization goals, and performance metrics that are used are what make an algorithmic system unique.When AI models are used in sustainability analytics, they usually look at how well they work, how well they compare to other models, or how accurate their predictions are. This could mean that environmental indicators that are easy to measure are given too much importance. These optimization methods can make environmental facts too simple to understand, which can lead to results that don't show the whole picture or are too positive about how well sustainability is working. When these results are turned into stories about sustainability, this kind of greenwashing is more likely to happen.

5.3 Quality, Access to Data, and Prejudice

Automated greenwashing happens a lot because of things that have to do with data.AI systems use both historical and real-time data sets. These datasets might not be complete, be broken up, or be biased toward certain businesses, areas, or environmental factors. It might be hard to include effects on the environment that are hard to measure or value, while simple data points might get too much attention. These gaps in data can make algorithmic results look better than they really are or hide bad effects.

5.4 Gaps in the Law and Government

It's getting clearer how to talk about ESG and AI governance, but the rules haven't kept up with how these systems are used to market sustainability. There aren't clear rules in a lot of places about how algorithms should be described, clear, or responsible in ESG reports.If there isn't strong oversight, they might not do enough research before using their own AI tools or third-party ESG rating systems.This can lead to algorithmic greenwashing that lasts.

5.5 Market Incentives and the Role of Third Parties

More middlemen have entered the market for ESG ratings, sustainability analytics, and AI-powered reporting tools as it has grown.Third-party providers often have their own algorithms that change how success in sustainability is tracked and shared. People in this market might feel like they have to use simple scores or ranks that make everyone happy but don't show how they were made. This reliance on outside algorithmic authority can make greenwashing more likely and put a lot of people at risk.

5.6 People Don't Know Enough About AI or How to Use It

It makes algorithmic greenwashing even worse that managers, sustainability workers, and stakeholders don't know much about AI. If the people in charge don't know how to think critically about what computers say, they might believe AI's claims about sustainability without looking into them enough.If there isn't enough human oversight, the same thing can happen with official messages about sustainability. This makes it even more risky to trust.

When strategic purpose, technological design, data limitations, and institutional environments all come together, these things show that algorithmic greenwashing takes place. Fixing the technology problems isn't enough to make it work. The government needs to step in and make sure that AI is used in a way that is consistent with real goals for sustainability. We also need businesses to know about the issue.6. What does this mean for trust between the parties?

Customers are less likely to trust businesses that use AI for sustainability marketing and ESG reporting. This is especially true when AI is used to make things look better than they really are. People don't believe claims of sustainability very much because the information isn't always right, it's hard to do, and the effects on the environment last a long time.As algorithms play a bigger role in shaping stories about sustainability, false or exaggerated claims hurt more than just people's reputations; they also hurt the structures of authority and government.

One thing that will happen right away is that AI-generated messages about sustainability will no longer be trusted. People who care about the issue often think that algorithmic results are fair and based on facts.They also believe claims about sustainability made by AI more than those made by traditional narrative reports. People might quickly lose faith in both the group and the AI systems that run it if they later question these claims or see that they aren't true in real life. People may be less likely to believe future claims about sustainability if they lose trust in the person making them, even if they are true.

When institutions use algorithms to greenwash, they are less likely to be trusted in terms of ESG standards and reporting systems. Using AI-powered tools and third-party ESG ratings to decide on sustainability are becoming more important. However, if algorithms give doubtful results, it could damage trust in the ESG community as a whole. Because of this, important stakeholders and groups might not take ESG scores, benchmarks, and AI-powered analytics as seriously as they should. This would make them less useful for running the country.

The effects on getting people involved are also very important.For deep conversations, working together, and committing to long-term projects that will help the environment, you need to trust each other. If stakeholders think that algorithms are controlling sustainability communication or that it isn't clear enough, they might stop participating. They might also doubt what the group says or ask for better checks and balances.

Another important effect has to do with how responsibilities are divided. When algorithmic greenwashing happens, it's hard to know who is to blame. It could be people who work for the company, AI developers, data providers, or rating bodies from outside the company.People might not trust them as much if they aren't clear about ethical ownership and control.AI is far from perfect and experts and techinicals who are using the AI systems should understand pitfalls of AI.Lastly, algorithmic greenwashing makes people less likely to trust all of Green AI, not just some companies.As more people learn about the dangers of AI, repeated examples of AI giving false information about sustainability may make more people question its use in environmental governance.This question could make people less likely to use useful Green AI apps, which would make it harder to reach sustainability goals. In short, algorithmic greenwashing is a big problem for people's trust in groups, institutions, and the system as a whole. More powerful regulatory frameworks are required in place that stress transparency, responsibility, and human oversight in AI-based sustainable communication to fix these trust issues.

6. MANAGING THE RISKS: TRANSPARENCY AND GOVERNANCE MECHANISMS

To address the potential risks of algorithmic greenwashing, it's time to stop thinking and considering Green AI merely a technical issue or a technical process. its more than that. Companies must recognize it as a sociotechnical ecosystem that is deeply rooted in organizations. Transparency, accountability, and governance systems that make sure AI-enabled sustainable communication is in line with real environmental performance rather than simply symbolic disclosure are required to reduce risks adequately. Effective risk mitigation strategies must be developed to enhance trust.

6.1 Explainable and Transparent AI Systems

Explanability is the cornerstone of trust in AI powered ESG reporting.If trust is not built, then its is just waste of time. Explainable AI (XAI) strategies empower businessess who have a stake in the process to understand how environmental data is collected,processed and refined into ESG scores or sustainability claims. By making algorithmic logic clearer, firms can lower the chance that opaque models will hide biased assumptions or selective signs. Transparency in AI systems requires robust evaluation internally and externally.. This makes sustainability reports more accountable and reliant from stakeholders perspective.

6.2 Robust Data Governance and Quality Assurance

Organisations should establish clear standards for data collection, validation, and updating to ensure that AI systems reflect accurate and representative environmental performance .Data governance mechanisms should address issues of data bias, completeness, and relevance, particularly for environmental impacts that are difficult to quantify. Regular audits of data sources and methodologies can help prevent the systematic misrepresentation of sustainability outcomes. Data governance frameworks must address all the concerns related to bias, relevance, and inconsistencies. Regular audits must be conducted to make sure there are no issues and in case of any issues, they have been addressed.

6.3 Human-in-the-Loop Oversight

Human oversight remains important in mitigating algorithmic greenwashing risks. Human-in-the-loop models ensure that AI-generated sustainability outputs are reviewed, contextualised, and validated by qualified professionals before being communicated to stakeholders. Such oversight helps prevent the uncritical adoption of algorithmic outputs and reinforces ethical responsibility by maintaining clear lines of accountability for sustainability claims. Human intervention ensures that outputs are adequately monitored by qualified professionals.Human oversight is more reliable and trustworthy.AI driven processess with human judgment ensure to eliminate bias, clear alignment to the goals and objectives of the organization, and reinforcing ethical responsibility.

6.4. Guidelines for ethical AI and how organizations should run them

To make sure that Green AI is in line with sustainability goals, organizations need to include ethical concerns in their AI governance procedures. Fairness, responsibility, openness, and protecting the environment must all be clearly stated in ethical AI principles. Algorithm biases must be corrected, and organizations must develop instruments that are intended to detect bias and to reduce it as much as possible. The self-regulation mechanism is scalable and fosters technological innovation.

6.5 Standards for External Assurance and Regulation

Regulatory action is very important to stop algorithmic greenwashing on a large scale. Setting clear rules for AI disclosure and explainability, and responsibility in ESG reporting can help make things clearer and make sure that all businesses follow the same rules. External assurance measures, such as impartial audits of AI-driven disclosures related to sustainability and ESG data, can enhance credibility and trust among stakeholders through objective and external validation. External audits enhance the credibility of the ESG reporting.

6.6 Getting Stakeholders Involved and Interested

Finally, to deal with the possible risks of algorithmic greenwashing, stakeholders must be be involved, and AI literacy needs to be improved. Engaging stakeholders with what AI systems can do can help them better understand claims about sustainability. Detailed information on the utilization of AI in sustainability reporting can aid in establishing appropriate standards and mitigating excessive reliance on algorithmic authority. These frameworks for openness and governance show that dealing with algorithmic greenwashing isn't just a tech problem; it's also a management problem.

7. THE ROAD AHEAD: POLICY, MANAGERIAL, AND RESEARCH DIRECTIONS

As Green AI continues to evolve and shape the sustainability reporting and ESG communication, the major risks evolving from algorithmic greenwashing require joint efforts from all the disciplines in an organization, regulatory frameworks, and academic research. In the future, GREEN AI will become more prevalent and applicable in all organizations so it will be critical to develop strong frameworks, rigorous audits plus extensive support systems. The Policy framework dealing with addressing algorithmic transparency, explainability, and accountability should be further strengthened.

Policymakers must consider mandatory independent audits and checks of AI-enabled sustainability analytics and ESG scoring systems. These audits not only ensure data accuracy but also detect algorithmic bias, model robustness, and alignment with sustainability objectives. International coordination among regulators is particularly important, given the global nature of ESG markets and the cross-border use of AI-driven reporting platforms.

7.1 Managerial Directions

Organisations need to develop strong internal capabilities to critically evaluate AI-generated sustainability results rather than depending entirely on automated systems. This includes investing in AI literacy, cross-functional collaboration between teams of sustainability and compliance teams, and clear accountability structures for sustainability communication. Managers should prioritise transparency in the reporting of sustainability. Clearly communicating the role, benefits, challenges, and potential problems of AI systems can help manage stakeholder expectations and reduce too much dependence on AI systems.

The integration of human processes and oversight, balanced by fair ethical review processes into AI must be taken into account because it can further strengthen trust and eliminate the misuse of algorithmic greenwashing practices.

7.2 Research Directions

Strong research in the future is needed to develop validated frameworks and systems for assessing transparency, accountability, and trust in AI-enabled sustainability communication. Also, combining insights from information systems, sustainability studies, ethics, and marketing can advance conceptual clarity and methodological rigor. It is not advisable to rely just on one source of information. Longitudinal studies may also bring clarity on how repeated exposure to AI-mediated sustainability narratives influences stakeholder trust over time. Comparative research from various sources may result in positive outcomes across regions with differing levels of AI governance and would also be valuable.

7.3 Toward Responsible Green AI

AI has the potential to improve sustainability decision-making and communication significantly, only if its use and application are accompanied by robust governance, ethical oversight, and stakeholder engagement and involvement. The algorithmic greenwashing concerns require attention and understanding that trust is not only a product of technology, but it is an outcome of good accountability, transparency, elimination of bias, and a strong ethos system in companies. By aligning policy frameworks, managerial strategies, and research agendas, stakeholders can ensure that Green AI contributes to genuine sustainability outcomes rather than reinforcing symbolic or misleading narratives. Such alignment is essential for protecting stakeholder trust and realizing the transformative potential of Green AI in advancing environmental responsibility.

8. CONCLUSION

The increasing adoption of Green AI in sustainability reporting and ESG communication has revamped how organizations are reporting environmental information. AI-enabled systems have significant potential to enhance efficiency, analytical depth, and scalability. This chapter has examined how companies also introduce new and under-examined risks for stakeholder trust. Precisely, the emergence of algorithmic greenwashing represents a critical challenge at the intersection of technology, sustainability governance, and ethical accountability. Green AI is more environmentally friendly and significant in terms of results(Bolón-Canedo et al., 2024)

Algorithmic greenwashing as a technology-enabled extension of traditional greenwashing, this chapter has highlighted how opaque algorithms, biased data, and over-reliance on algorithmic authority can unintentionally or strategically produce misleading sustainability narratives. The analysis further underscores that stakeholder trust in the age of AI is shaped not only by the accuracy of sustainability claims but also by transparency, accountability, and the presence of meaningful human oversight. The major focus of this chapter insisted on the designing of robust governance frameworks, explainable AI, and ethical oversight can help restore confidence and legitimacy in Green AI applications.

The future demands more responsible deployment of Green AI which requires a deliberate and coordinated approach that integrates policy intervention, managerial responsibility, and continued scholarly intervention. Algorithmic greenwashing is not simply a technical challenge but a governance imperative central to the credibility of sustainability communication and the long-term viability of ESG frameworks.

In conclusion, the future of Green AI will be defined not by its computational capabilities alone but by the extent to which it is governed in ways that reinforce trust, legitimacy, and genuine sustainability outcomes. Recognizing and addressing the risks of algorithmic greenwashing is therefore essential to navigating the challenges and the road ahead.

REFERENCES

Alzoubi, Y. I., & Mishra, A. (2024). Green artificial intelligence initiatives: Potentials and challenges. *Journal of Cleaner Production*, *468*, 143090. DOI: 10.1016/j.jclepro.2024.143090

Biggi, G., Iori, M., Mazzei, J., & Mina, A. (2025). Green intelligence: The AI content of green technologies. *Eurasian Business Review*, *15*(3), 803–840. DOI: 10.1007/s40821-024-00288-1

Bolón-Canedo, V., Morán-Fernández, L., Cancela, B., & Alonso-Betanzos, A. (2024). A review of green artificial intelligence: Towards a more sustainable future. *Neurocomputing*, *599*, 128096. DOI: 10.1016/j.neucom.2024.128096

Chaudhary, G. (2024). Unveiling the black box: Bringing algorithmic transparency to AI. *Masaryk University Journal of Law and Technology*, *18*(1), 93–122. DOI: 10.5817/MUJLT2024-1-4

Dash, S. (2025). Green AI: Enhancing sustainability and energy efficiency in AI-integrated enterprise systems. *IEEE Access : Practical Innovations, Open Solutions*, *13*, 21216–21228. DOI: 10.1109/ACCESS.2025.3532838

Eze, M. (2025). Digital greenwashing in the age of sustainability marketing: A meta-analysis of consumer perception, detection, and backlash. *International Journal of Marketing Studies*, *17*(2), 32. DOI: 10.5539/ijms.v17n2p32

Li, J., Wu, T., Hu, B., Pan, D., & Zhou, Y. (2025). Artificial intelligence and corporate ESG performance. *International Review of Financial Analysis*, *102*, 104036. DOI: 10.1016/j.irfa.2025.104036

Mehta, Y., Xu, R., Lim, B., Wu, J., & Gao, J. (2023). A review for green energy machine learning and AI services. *Energies*, *16*(15), 5718. DOI: 10.3390/en16155718

Shankar, R., & Xu, Q. (2025). Automated detection of greenwashing in Indian corporate sustainability reports using natural language processing. DOI: 10.2139/ssrn.5294480

Tabbakh, A., Al Amin, L., Islam, M., Mahmud, G. M., Chowdhury, I. K., & Mukta, M. S. (2024). Towards sustainable AI: A comprehensive framework for green AI. *Discover Sustainability*, *5*(1), 408. Advance online publication. DOI: 10.1007/s43621-024-00641-4

Von Struensee, S. (2021). Artificial intelligence and corporate social responsibility: Employees' key role in driving responsible artificial intelligence at big tech. SSRN *Electronic Journal*. DOI: 10.2139/ssrn.3873097

Wang, N., Pan, H., Feng, Y., & Du, S. (2023). How do ESG practices create value for businesses? Research review and prospects. *Sustainability Accounting. Management and Policy Journal, 15*(5), 1155–1177. DOI: 10.1108/SAMPJ-12-2021-0515

Wang, Q., Qi, Y., & Li, R. (2025). Artificial intelligence and corporate sustainability: Shaping the future of <scp>ESG</scp> in the age of industry 5.0. *Sustainable Development*. Advance online publication. DOI: 10.1002/sd.70200

Zaman, S. I., Jadoon, S. T., & Khan, S. A. (2024). Green algorithms: The impact of artificial intelligence on environmental sustainability. *1st International Conference on Industrial, Manufacturing, and Process Engineering (ICIMP-2024)*, 40. DOI: 10.3390/engproc2024076040

Compilation of References

Aarthi, S., Ravikumar, R. N., Balakrishnan, R., Pavithra, K., Jumaniyozov, F., & Praveen, V. M. (2025). Blockchain for seamless and transparent smart hospitality operations. In *Smart Operations and Enhancing Guest Experience in the Hospitality Industry* (pp. 187–218). IGI Global., DOI: 10.4018/979-8-3373-2145-5.ch007

Aarthi, S., Ravikumar, R. N., Sobirova, K., & Mamadiyarov, Z. (2025). Empowering a sustainable future through digital learning and the SDGs. In *Harnessing E-Learning to Create a Sustainable Future* (pp. 145–172). IGI Global Scientific Publishing., DOI: 10.4018/979-8-3373-0112-9.ch006

Abebaw, S. E. (2025). A Global Review of the Impacts of Climate Change and Variability on Agricultural Productivity and Farmers' Adaptation Strategies. *Food Science & Nutrition*, *13*(5), e70260. DOI: 10.1002/fsn3.70260 PMID: 40370417

Abid, S. K., Roosli, R., Nazir, U., & Kamarudin, N. S. (2025). AI-enhanced crowdsourcing for disaster management: Strengthening community resilience through social media. *International Journal of Emergency Medicine*, *18*(1), 201. DOI: 10.1186/s12245-025-01009-9 PMID: 41083937

Adil, L., Eckstein, D., Künzel, V., & Schäfer, L. (2025). Climate risk index 2025.

AfDB. (2020). *Climate Adaptation in African Agriculture: Digital and AI-Based Early Warning Systems*. African Development Bank.

Agenda, R. G. (2024). Regulatory Experimentation: Moving ahead on the Agile Regulatory Governance Agenda.

Agrawal, T., Kalla, A., & Acharya, B. (2021). Blockchain-based traceability in agriculture supply chains: A review. *Journal of Cleaner Production*, *298*, 126812.

Agyeman, J., Schlosberg, D., Craven, L., & Matthews, C. (2016). Trends and directions in environmental justice: From inequity to everyday life, community, and just sustainabilities. *Annual Review of Environment and Resources*, *41*(1), 321–340. DOI: 10.1146/annurev-environ-110615-090052

Ahmadi, A., Talaei, M., Sadipour, M., Amani, A. M., & Jalili, M. (2023). Deep federated learning-based privacy-preserving wind power forecasting. *IEEE Access : Practical Innovations, Open Solutions*, *11*, 39521–39530. DOI: 10.1109/ACCESS.2022.3232475

Ahmadi, M. H., Ghazvini, M., Sadeghzadeh, M., Nazari, M. A., Kumar, R., Naeimi, A., & Ming, T. (2022). Application of artificial intelligence for renewable energy systems: A review. *Renewable & Sustainable Energy Reviews*, *156*, 111821.

Ahmad, T., Zhang, D., Huang, C., Zhang, H., Dai, N., Song, Y., & Chen, H. (2021). Artificial intelligence in sustainable energy industry: Status quo, challenges and opportunities. *Journal of Cleaner Production*, *289*, 125834. DOI: 10.1016/j.jclepro.2021.125834

Ahmed, S., Hilal-Alnaqbi, A., Al Hemairy, M., & Al Ahmad, M. (2018). ECG abnormality detection algorithm. *International Journal of Advanced Computer Science and Applications*, *9*(8), 215–219. DOI: 10.14569/IJACSA.2018.090827

Akbar, N. S., Batool, G., Siddiqui, S., & Habib, M. B. (2025). Leveraging AI for Environmental Solutions addressing Climate Change, Pollution Monitoring and Sustainable Resource Management. *Materials Research Proceedings*, *48*, 893–902. DOI: 10.21741/9781644903414-97

Al Hemairy, M., Amin, S., Hijji, M., Serhani, M., & Al Ahmad, M. (2016). Integrated and scalable architecture for providing cost-effective remote health monitoring. In *9th International Conference on Developments in eSystems Engineering (DeSE)* (pp. 74–80). IEEE. DOI: 10.1109/DeSE.2016.14

Albahri, A. S., Khaleel, Y. L., Habeeb, M. A., Ismael, R. D., Hameed, Q. A., Deveci, M., Homod, R. Z., Albahri, O. S., Alamoodi, A. H., & Alzubaidi, L. (2024). A systematic review of trustworthy artificial intelligence applications in natural disasters. *Computers & Electrical Engineering*, *118*, 109409. DOI: 10.1016/j.compeleceng.2024.109409

Aldeghi, M., Gapsys, V., & de Groot, B. L. (2019). Predicting Kinase Inhibitor Resistance: Physics-Based and Data-Driven Approaches. *ACS Central Science*, *5*(8), 1468–1474. DOI: 10.1021/acscentsci.9b00590 PMID: 31482130

Alduais, Y., Zhang, H., Fan, F., Chen, J., & Chen, B. (2023). Non-small cell lung cancer (NSCLC): A review of risk factors, diagnosis, and treatment. *Medicine, 102*(8), e32899–e32899. DOI: 10.1097/MD.0000000000032899 PMID: 36827002

Ali, K., & Rogers, D. J. (2021). An orientation-independent multi-input energy harvesting wireless sensor node. IEEE Transactions on Industrial Electronics (1982), 68(2), 1665–1674. DOI: 10.1109/TIE.2020.2967719

Ali, D. M. T. E., Motuzienė, V., & Džiugaitė Tumėnienė, R. (2024). Ai driven innovations in building energy management systems: A review of potential applications and energy savings. *Energies, 17*(17), 4277. DOI: 10.3390/en17174277

Al-Jarf, R., de Sá, A. G., Pires, D. E., & Ascher, D. B. (2021). pdCSM-cancer: Using graph-based signatures to identify small molecules with anticancer properties. *Journal of Chemical Information and Modeling, 61*(7), 3314–3322. DOI: 10.1021/acs.jcim.1c00168 PMID: 34213323

Almond, R. E. A., Grooten, M., Juffe Bignoli, D., & Petersen, T. (2022). Living Planet Report 2022–Building a nature-positive society. *WWF: Gland, Switzerland.*

Alzoubi, Y. I., & Mishra, A. (2024). Green artificial intelligence initiatives: Potentials and challenges. *Journal of Cleaner Production, 468*, 143090. DOI: 10.1016/j.jclepro.2024.143090

Amani, M., Ghorbanian, A., Ahmadi, S. A., Kakooei, M., Moghimi, A., Mirmazloumi, S. M., Moghaddam, S. H. A., Mahdavi, S., Ghahremanloo, M., Parsian, S., Wu, Q., & Brisco, B. (2020). Google Earth Engine Cloud Computing Platform for Remote Sensing Big Data Applications: A Comprehensive Review. *IEEE Journal of Selected Topics in Applied Earth Observations and Remote Sensing, 13*, 5326–5350. DOI: 10.1109/JSTARS.2020.3021052

Bajaber, F. (2023). Adaptive density control based on random sensing range for energy efficiency in IoT sensor networks. *Pertanika Journal of Science & Technology, 31*(4). Advance online publication. DOI: 10.47836/pjst.31.4.14

Banner, R., Hubara, I., Hoffer, E., & Soudry, D. (2019). Post-training 4-bit quantization of convolution networks for rapid deployment. *Advances in Neural Information Processing Systems, 32*, 1–12.

Barrett, C. B., Benton, T. G., Cooper, K. A., Fanzo, J., & Gandhi, R.. (2020). Socio-technical innovation bundles for agri-food systems transformation. *Nature Sustainability, 3*, 103–110. DOI: 10.1038/s41893-020-00661-8

Bastiaanssen, W. G. M., Molden, D. J., & Makin, I. W. (2000). Remote sensing for irrigated agriculture: Examples from research and possible applications. *Agricultural Water Management*, *46*(2), 137–155. DOI: 10.1016/S0378-3774(00)00080-9

Batty, M., Axhausen, K. W., Giannotti, F., Pozdnoukhov, A., Bazzani, A., Wachowicz, M., Ouzounis, G., & Portugali, Y. (2012). Smart cities of the future. *The European Physical Journal. Special Topics*, *214*(1), 481–518. DOI: 10.1140/epjst/e2012-01703-3

Ben Bouallègue, Z., Clare, M., Magnusson, L., Gascón, E., Maier-Gerber, M., Janoušek, M., Rodwell, M., Pinault, F., Dramsch, J., Lang, S., Raoult, B., Rabier, F., Chevallier, M., Sandu, I., Düben, P., Chantry, M., & Pappenberger, F. (2024). The Rise of Data-Driven Weather Forecasting: A First Statistical Assessment of Machine Learning–Based Weather Forecasts in an Operational-Like Context. *Bulletin of the American Meteorological Society*, *105*(6), E864–E883. Advance online publication. DOI: 10.1175/BAMS-D-23-0162.1

Benos, L., Tagarakis, A. C., Dolias, G., Berruto, R., Emmanouilides, C., & Bochtis, D. (2021). Machine learning in agriculture: A comprehensive updated review. *Sensors (Basel)*, *21*(11), 3758. DOI: 10.3390/s21113758 PMID: 34071553

Berente, N., Gu, B., Recker, J., & Santhanam, R. (2021). Managing Artificial Intelligence. *Management Information Systems Quarterly*, *45*(3), 1433–1450. DOI: 10.25300/MISQ/2021/16274

Berg, F., Koelbel, J. F., & Rigobon, R. (2022). Aggregate confusion: The divergence of ESG ratings. *Review of Finance*, *26*(6), 1315–1344. DOI: 10.1093/rof/rfac033

Bhullar, K. S., Lagarón, N. O., McGowan, E. M., Parmar, I., Jha, A., Hubbard, B. P., & Rupasinghe, H. V. (2018). Kinase-targeted cancer therapies: Progress, challenges and future directions. *Molecular Cancer*, *17*(1), 1–20. DOI: 10.1186/s12943-018-0804-2 PMID: 29455673

Bibri, S. E., & Krogstie, J. (2017). Smart Sustainable Cities of the Future: An Extensive Interdisciplinary Literature Review. *Sustainable Cities and Society*, *31*, 183–212. DOI: 10.1016/j.scs.2017.02.016

Biggi, G., Iori, M., Mazzei, J., & Mina, A. (2025). Green intelligence: The AI content of green technologies. *Eurasian Business Review*, *15*(3), 803–840. DOI: 10.1007/s40821-024-00288-1

Bi, K., Xie, L., Zhang, H., Chen, X., Gu, X., & Tian, Q. (2023). Accurate medium-range global weather forecasting with 3D neural networks. *Nature*, *619*(7970), 533–538. DOI: 10.1038/s41586-023-06185-3 PMID: 37407823

Bi, K., Zhou, Y., Lin, Y., Wang, J., Zhang, H., & Liu, Y. (2023). *Pangu-Weather: A 3D high-resolution model for fast and accurate global weather forecasting*. arXiv. https://arxiv.org/abs/2211.02556

Bolón-Canedo, V., Morán-Fernández, L., Cancela, B., & Alonso-Betanzos, A. (2024). A review of green artificial intelligence: Towards a more sustainable future. *Neurocomputing*, *599*, 128096. DOI: 10.1016/j.neucom.2024.128096

Brenowitz, N. D., & Bretherton, C. S. (2018). Prognostic validation of a neural network unified physics parameterization. *Geophysical Research Letters*, *45*(12), 6289–6298. DOI: 10.1029/2018GL078510

Bronson, K., & Knezevic, I. (2016). Big Data in food and agriculture. *Big Data & Society*, *3*(1), 1–5. DOI: 10.1177/2053951716648174

Bukowski, L., & Werbinska-Wojciechowska, S. (2025). Towards maintenance 5.0: Resilience-based maintenance in AI-driven sustainable and human-centric industrial systems. *Sensors (Basel)*, *25*(16), 5100. DOI: 10.3390/s25165100 PMID: 40871964

Burki, A. K., Ahamed Mafaz, M. N., Ahmad, Z., Zulfaka, A., & Bin Isa, M. Y. (2024). Artificial Intelligence and environmental sustainability: Insights from PLS-SEM on resource efficiency and carbon emission reduction. OPSearch. *American Journal of Operations Research*, *3*(10), 277–288. DOI: 10.58811/opsearch.v3i10.141

Busch, T., Johnson, M., & Pioch, T. (2022). Corporate carbon performance data: Quo vadis? *Journal of Industrial Ecology*, *26*(1), 350–363. DOI: 10.1111/jiec.13008

Camidge, D. R., Pao, W., & Sequist, L. V. (2014). Acquired resistance to TKIs in solid tumours: Learning from lung cancer. In *Nature Reviews Clinical Oncology* (Vol. 11, Issue 8, pp. 473–481). Nature Publishing Group. DOI: 10.1038/nrclinonc.2014.104

Camps-Valls, G., Fernández-Torres, M. Á., Cohrs, K. H., Höhl, A., Castelletti, A., Pacal, A., Robin, C., Martinuzzi, F., Papoutsis, I., Prapas, I., Pérez-Aracil, J., Weigel, K., Gonzalez-Calabuig, M., Reichstein, M., Rabel, M., Giuliani, M., Mahecha, M. D., Popescu, O. I., Pellicer-Valero, O. J., & Williams, T. (2025). Artificial intelligence for modeling and understanding extreme weather and climate events. *Nature Communications*, *16*(1), 1919. Advance online publication. DOI: 10.1038/s41467-025-56573-8 PMID: 39994190

Capetillo-Contreras, O., Rodríguez-Pulido, J. A., Hernández-Aguilar, M., & Parra-Flores, J. (2024). Artificial intelligence-based aquaculture systems: Monitoring, management, and sustainability perspectives. *Journal of Marine Science and Engineering*, *12*(1), 161. DOI: 10.3390/jmse12010161

Chaudhary, G. (2024). Unveiling the black box: Bringing algorithmic transparency to AI. *Masaryk University Journal of Law and Technology*, *18*(1), 93–122. DOI: 10.5817/MUJLT2024-1-4

Chen, L., Zhong, X., Zhang, F., Cheng, Y., Xu, Y., Qi, Y., & Li, H. (2023). FuXi: A cascade machine learning forecasting system for 15-day global weather forecast. *NPJ Climate and Atmospheric Science*, *6*(1), 190. DOI: 10.1038/s41612-023-00512-1

Chen, S.-C., Chen, H.-M., Chen, H.-K., & Li, C.-L. (2024). Multi-objective optimization in Industry 5.0: Human-centric AI integration for sustainable and intelligent manufacturing. *Processes (Basel, Switzerland)*, *12*(12), 2723. DOI: 10.3390/pr12122723

Chen, T. (2016). XGBoost: A Scalable Tree-Boosting System. *Proceedings of the 22nd ACM SIGKDD International Conference on Knowledge Discovery and Data Mining*, 785–794.

Cheong, S. M., Sankaran, K., & Bastani, H. (2022). Artificial intelligence for climate change adaptation. *Wiley Interdisciplinary Reviews. Data Mining and Knowledge Discovery*, *12*(5), 1–7. DOI: 10.1002/widm.1459

Choi, Y., & Chan, A. P. (2015). PROVEAN web server: A tool to predict the functional effect of amino acid substitutions and indels. *Bioinformatics (Oxford, England)*, *31*(16), 2745–2747. DOI: 10.1093/bioinformatics/btv195 PMID: 25851949

Christin, S., Hervet, É., & Lecomte, N. (2019). Applications for deep learning in ecology. *Methods in Ecology and Evolution*, *10*(10), 1632–1644. DOI: 10.1111/2041-210X.13256

Corbane, C., Syrris, V., Seliger, R., Bjørnerud, M., & Kemper, T. (2021). Remote sensing for biodiversity monitoring: A review of trends and future directions. *Remote Sensing*, *13*(18), 3654.

Crane-Droesch, A. (2018). Machine learning methods for crop yield prediction and climate risk assessment in agriculture. *Environmental Research Letters*, *13*(11), 114003. DOI: 10.1088/1748-9326/aae159

Dash, S. (2025). Green AI: Enhancing sustainability and energy efficiency in AI-integrated enterprise systems. *IEEE Access : Practical Innovations, Open Solutions*, *13*, 21216–21228. DOI: 10.1109/ACCESS.2025.3532838

Delmas, M. A., & Burbano, V. C. (2011). The drivers of greenwashing. *California Management Review*, *54*(1), 64–87. DOI: 10.1525/cmr.2011.54.1.64

Eda, T., Kurebayashi, R., Shi, X., Enomoto, S., Iida, K., & Hamuro, D. (2021). An efficient event-driven inference approach to support AI applications in IOWN era. *NTT Technical Review*, *19*(2), 41–46. DOI: 10.53829/ntr202102fa4

Elhady, A. M., & Shohieb, S. (2025). AI-driven sustainable finance: Computational tools, ESG metrics, and global implementation. *Future Business Journal*, *11*(1), 209. Advance online publication. DOI: 10.1186/s43093-025-00610-x

Entwistle, A. (1988). What Is Artificial Intelligence? *Engineering Materials and Design*, *32*(3), 1–14. DOI: 10.55248/gengpi.2022.31261

European Commission. (2020). [*Project Results: Digital Technologies for Biodiversity Monitoring*. Publications Office of the European Union.]. *Horizon*, •••, 2020.

European Commission. (2022). *Artificial intelligence and climate change*. Publications Office of the European Union.

European Environment Agency (EEA). (2021). *Biodiversity Monitoring Using Remote Sensing and UAVs in Agricultural Landscapes*. EEA Technical Report.

European Union. (2020). *The European Green Deal and Common Agricultural Policy (CAP) Biodiversity Strategy 2030*. EU Publications.

Evenson, R. E., & Gollin, D. (2003). Assessing the Impact of the Green Revolution, 1960–2000. *Science*, *300*(5620), 758–762. DOI: 10.1126/science.1078710 PMID: 12730592

Eze, M. (2025). Digital greenwashing in the age of sustainability marketing: A meta-analysis of consumer perception, detection, and backlash. *International Journal of Marketing Studies*, *17*(2), 32. DOI: 10.5539/ijms.v17n2p32

Falk, S., & van Wynsberghe, A. (2024). Challenging AI for Sustainability: What ought it mean? *AI and Ethics*, *4*(4), 1345–1355. DOI: 10.1007/s43681-023-00323-3

FAO. (2013). *Climate-Smart Agriculture Sourcebook*. Food and Agriculture Organization of the United Nations.

FAO. (2017). *The future of food and agriculture: Trends and challenges*. Food and Agriculture Organization of the United Nations.

FAO. (2018). *The State of Food Security and Nutrition in the World 2018*. Food and Agriculture Organization of the United Nations.

FAO. (2019). *The State of the World's Biodiversity for Food and Agriculture*. Food and Agriculture Organization of the United Nations.

FAO. (2021). [*Making agrifood systems more resilient to shocks and stresses.* Food and Agriculture Organization of the United Nations.]. *The State of Food and Agriculture*, •••, 2021.

FAO. (2021). *The State of Food Security and Nutrition in the World 2021*. Food and Agriculture Organization of the United Nations.

FAO. (2022). *Digital Agriculture and Inclusive Growth: Bridging the Data Divide*. Food and Agriculture Organization of the United Nations.

Ferrag, M. A., Shu, L., Yang, X., Derhab, A., & Maglaras, L. (2020). Smart agriculture security: A survey on IoT security for smart farming. *IEEE Access : Practical Innovations, Open Solutions*, *8*, 145000–145019.

Fletcher, S., March, A. L. A., Roberts, K., Shirian, Y., i Canals, L. M., Cairns, A., ... & Bass, A. (2023). *Turning off the Tap: How the world can end plastic pollution and create a circular economy*. United Nations Environment Programme.

Floridi, L., & Cowls, J. (2022). A unified framework of five principles for AI in society. *Machine learning and the city: Applications in architecture and urban design*, 535-545.

Gagic, Z., Ruzic, D., Djokovic, N., Djikic, T., & Nikolic, K. (2020). In silico methods for design of kinase inhibitors as anticancer drugs. *Frontiers in Chemistry*, *7*, 873. DOI: 10.3389/fchem.2019.00873 PMID: 31970149

Geng, C., Vangone, A., Folkers, G. E., Xue, L. C., & Bonvin, A. M. (2019). iSEE: Interface structure, evolution, and energy-based machine learning predictor of binding affinity changes upon mutations. *Proteins*, *87*(2), 110–119. DOI: 10.1002/prot.25630 PMID: 30417935

Golian, S., Murphy, C., & Meresa, H. (2021). Regionalization of hydrological models for flow estimation in ungauged catchments in Ireland. *Journal of Hydrology. Regional Studies*, *36*(100859), 100859. DOI: 10.1016/j.ejrh.2021.100859

GreenFi. (2024). *Case study: GreenFi ESG AI for asset emissions (UOB Bank)*. GreenFi Case Studies.

GreenFi. (2025). *IFRS S2: Strengthening physical climate risk reporting*. GreenFi Insights.

Guei, A. C., Christin, S., Lecomte, N., & Hervet, É. (2024). ECOGEN: Bird sounds generation using deep learning. *Methods in Ecology and Evolution*, *15*(1), 69–79. DOI: 10.1111/2041-210X.14239

Guo, W. (2024). Overview of research progress and challenges in federated learning. *Transactions on Computer Science and Intelligent Systems Research*, *5*, 797–804. DOI: 10.62051/9qyaha16

Gupta, P., & Singh, R. (2023). Green computing strategies for sustainable artificial intelligence applications. *Journal of Sustainable Technology and Innovation*, *12*(3), 155–172.

Gu, S., Liu, H., Liu, L., Hou, T., & Kang, Y. (2023). Artificial intelligence methods in kinase target profiling: Advances and challenges. *Drug Discovery Today*, *28*(11), 103796. DOI: 10.1016/j.drudis.2023.103796 PMID: 37805065

Gutiérrez, M., Moraga, M. Á., García, F., & Calero, C. (2024). Green IN Artificial Intelligence from a Software Perspective: State-of-the-Art and Green Decalogue. *ACM Computing Surveys*, *57*(3), 1–30. Advance online publication. DOI: 10.1145/3698111

Hakami, M. A. (2024). Harnessing machine learning potential for personalised drug design and overcoming drug resistance. *Journal of Drug Targeting*, *32*(8), 918–930. DOI: 10.1080/1061186X.2024.2365934 PMID: 38842417

Han, H., Liu, Z., Li, J., & Zeng, Z. (2024). Challenges in remote sensing based climate and crop monitoring: Navigating the complexities using AI. *Journal of Cloud Computing (Heidelberg, Germany)*, *13*(1), 34. Advance online publication. DOI: 10.1186/s13677-023-00583-8

Han, S., Pool, J., Tran, J., & Dally, W. J. (2015). Learning both weights and connections for efficient neural networks. *Advances in Neural Information Processing Systems*, *28*, 1135–1143.

Heo, S., Park, S., & Lee, D. K. (2023). Multi-hazard exposure mapping under climate crisis using random forest algorithm for the Kalimantan Islands, Indonesia. *Scientific Reports*, *13*(1), 1–13. DOI: 10.1038/s41598-023-40106-8 PMID: 37596300

Hinton, G., Vinyals, O., & Dean, J. (2015). *Distilling the knowledge in a neural network*. arXiv. https://arxiv.org/abs/1503.02531

Hooker, S., Courville, A., Clark, G., Dauphin, Y., & Frome, A. (2021). What do compressed deep neural networks forget? *Nature Machine Intelligence*, *3*(6), 438–446. DOI: 10.1038/s42256-021-00343-7

Howson, P., & de Vries, A. (2022). Preying on the poor? Opportunities and risks of blockchain technologies for climate finance. *Energy Research & Social Science*, *84*, 102394. DOI: 10.1016/j.erss.2021.102394

IBM News. (2025). *IBM and NASA Release Groundbreaking Open-Source AI Model on Hugging Face to Predict Solar Weather and Help Protect Critical Technology.*

ICRISAT. (2019). *AI-based Sowing and Irrigation Advisory Services for Smallholder Farmers*. International Crops Research Institute for the Semi-Arid Tropics.

International Energy Agency (IEA). (2023). *Digitalization and energy*. IEA Publishing.

International Renewable Energy Agency IRENA. (2022). *Renewable energy statistics 2022*. International Renewable Energy Agency IRENA.

IPBES. (2019). *Global Assessment Report on Biodiversity and Ecosystem Services*. Intergovernmental Science-Policy Platform on Biodiversity and Ecosystem Services.

IPCC. (2022). [*Impacts, Adaptation and Vulnerability.* Intergovernmental Panel on Climate Change.]. *Climatic Change*, •••, 2022.

IPCC. (2022). Climate Change 2022_Impacts, Adaptation and Vulnerability_Working Group II contribution to the Sixth Assessment Report of the Intergovernamental Panel on Climate Change. In *Working Group II contribution to the Sixth Assessment Report of the Intergovernamental Panel on Climate Change* (Issue August).

Jacob, B., Kligys, S., Chen, B., Zhu, M., Tang, M., Howard, A., Adam, H., & Kalenichenko, D. (2018). Quantization and training of neural networks for efficient integer-arithmetic-only inference. In *Proceedings of the IEEE Conference on Computer Vision and Pattern Recognition (CVPR)* (pp. 2704–2713). DOI: 10.1109/CVPR.2018.00286

Jayaram, A., & Dhanya, C. T. (2021). IoT-based smart agriculture: A review on enabling technologies and challenges. *Environmental Monitoring and Assessment, 193*(10), 1–20.

Johnstone, I.Organisation for Economic Co-operation and Development (OECD). (2022). 10. Organisation for Economic Co-operation and Development (OECD). *Yearbook of International Environmental Law, 33*(1), 271–276. DOI: 10.1093/yiel/yvad054

Jubb, H. C., Higueruelo, A. P., Ochoa-Montaño, B., Pitt, W. R., Ascher, D. B., & Blundell, T. L. (2017). Arpeggio: A web server for calculating and visualising interatomic interactions in protein structures. *Journal of Molecular Biology, 429*(3), 365–371. DOI: 10.1016/j.jmb.2016.12.004 PMID: 27964945

Kaack, L. H., Donti, P. L., Strubell, E., Kamiya, G., Creutzig, F., & Rolnick, D. (2022). Aligning artificial intelligence with climate change mitigation. *Nature Climate Change*, *12*(6), 518–527. DOI: 10.1038/s41558-022-01377-7

Kaine, G., & Wright, V. (2025). Regulator theory, natural hazards, and climate change. *Sustainability (Basel)*, *17*(7), 2979. DOI: 10.3390/su17072979

Kamble, S. S., Gunasekaran, A., & Arha, H. (2020). Understanding the blockchain technology adoption in supply chains. *International Journal of Production Research*, *58*(7), 2009–2033. DOI: 10.1080/00207543.2018.1518610

Kamilaris, A., & Prenafeta-Boldú, F. X. (2018). Deep learning in agriculture: A survey. *Computers and Electronics in Agriculture*, *147*, 70–90. DOI: 10.1016/j.compag.2018.02.016

Kermani, A., Zeraatkar, E., & Irani, H. (2025). *Energy-efficient transformer inference: Optimization strategies for time series classification*. arXiv. https://arxiv.org/abs/2502.16627

Kouhizadeh, M., Sarkis, J., & Zhu, Q. (2021). At the nexus of blockchain technology, the circular economy, and product deletion. *Transportation Research Part E, Logistics and Transportation Review*, *145*, 102153.

Kratzert, F., Klotz, D., Shalev, G., Klambauer, G., Hochreiter, S., & Nearing, G. (2019). Towards learning universal, regional, and local hydrological behaviors via machine learning applied to large-sample datasets. *Hydrology and Earth System Sciences*, *23*(12), 5089–5110. DOI: 10.5194/hess-23-5089-2019

Kshetri, N. (2021). Artificial intelligence, blockchain, and trust. *Computer*, *54*(1), 88–92.

Kumar, A. B., Sanjaya, K., & Sanjaya, K. (2025). AI renaissance in renewable energy. In *Advances in Environmental Engineering and Green Technologies* (pp. 15–32). IGI Global., DOI: 10.4018/979-8-3373-0045-0.ch002

Kumar, R., Chaudhary, K., Gupta, S., Singh, H., Kumar, S., Gautam, A., Kapoor, P., & Raghava, G. P. (2013). CancerDR: Cancer drug resistance database. *Scientific Reports*, *3*(1), 1445. DOI: 10.1038/srep01445 PMID: 23486013

Kurth, T., Subramanian, S., Harrington, P., Pathak, J., Mardani, M., Hall, D., Miele, A., Kashinath, K., & Anandkumar, A. (2023). FourCastNet: Accelerating Global High-Resolution Weather Forecasting Using Adaptive Fourier Neural Operators. *Proceedings of the Platform for Advanced Scientific Computing Conference*. DOI: 10.1145/3592979.3593412

Lal, R. (2020). Regenerative agriculture for food and climate. *Journal of Soil and Water Conservation*, *75*(5), 123A–124A. DOI: 10.2489/jswc.2020.0620A

Lam, R., Sanchez-Gonzalez, A., Willson, M., Wirnsberger, P., Fortunato, M., Alet, F., Ravuri, S., Ewalds, T., Eaton-Rosen, Z., Hu, W., Merose, A., Hoyer, S., Holland, G., Vinyals, O., Stott, J., Pritzel, A., Mohamed, S., & Battaglia, P. (2023). Learning skillful medium-range global weather forecasting. *Science*, *382*(6677), 1416–1421. DOI: 10.1126/science.adi2336 PMID: 37962497

Lapp, S., Rhinehart, T., Freeland-Haynes, L., Khilnani, J., Syunkova, A., & Kitzes, J. (2023). OpenSoundscape: An open-source bioacoustics analysis package for Python. *Methods in Ecology and Evolution*, *14*(9), 2321–2328. DOI: 10.1111/2041-210X.14196

Liakos, K. G., Busato, P., Moshou, D., Pearson, S., & Bochtis, D. (2018). Machine Learning in Agriculture: A Review. *Sensors (Basel)*, *18*(8), 2674. DOI: 10.3390/s18082674 PMID: 30110960

Li, J., Wu, T., Hu, B., Pan, D., & Zhou, Y. (2025). Artificial intelligence and corporate ESG performance. *International Review of Financial Analysis*, *102*, 104036. DOI: 10.1016/j.irfa.2025.104036

Lin, J., Chen, W. M., Cuvellier, L., Zhu, M., Lin, S., & Wagner, A. (2017). Adaptive neural networks for energy-constrained inference. In *Advances in Neural Information Processing Systems*.

Lin, R., Wang, X., & Chen, W. (2021). *Atmospheric Environment Studies*.

Linardos, V., Drakaki, M., Tzionas, P., & Karnavas, Y. L. (2022). Machine Learning in Disaster Management: Recent Developments in Methods and Applications. In *Machine Learning and Knowledge Extraction* (Vol. 4, Issue 2, pp. 446–473). DOI: 10.3390/make4020020

Li, T., Sahu, A. K., Talwalkar, A., & Smith, V. (2020). Federated learning: Challenges, methods, and future directions. *IEEE Signal Processing Magazine*, *37*(3), 50–60. DOI: 10.1109/MSP.2020.2975749

Liu, C., Yang, J., Lu, Z., Chen, C., Wang, J., Xu, D., & Li, X. (2023). Design and implementation of an event-driven smart sensor node for wireless monitoring systems. *Sensors (Basel)*, *23*(24), 9737. Advance online publication. DOI: 10.3390/s23249737 PMID: 38139583

Li, Z., Shahidehpour, M., Alabdulwahab, A., & Abusorrah, A. (2021). Bilevel model for power system operation with demand response and energy storage. *IEEE Transactions on Smart Grid*, *12*(4), 2921–2932.

Lynch, P. (2008). The origins of computer weather prediction and climate modeling. *Journal of Computational Physics, 227*(7), 3431–3444. DOI: 10.1016/j.jcp.2007.02.034

Lyon, T. P., & Montgomery, A. W. (2015). The means and end of greenwash. *Organization & Environment, 28*(2), 223–249. DOI: 10.1177/1086026615575332

Macarthur, E. L. L. E. N., & Heading, H. E. A. D. I. N. G. (2019). How the circular economy tackles climate change. *Ellen MacArthur Found, 1*, 1 71.

Malik, A., Saggi, M. K., Rehman, S., Sajjad, H., Inyurt, S., Bhatia, A. S., Farooque, A. A., Oudah, A. Y., & Yaseen, Z. M. (2022). Deep learning versus gradient boosting machine for pan evaporation prediction. *Engineering Applications of Computational Fluid Mechanics, 16*(1), 570–587. DOI: 10.1080/19942060.2022.2027273

Marisamy, K., Selvakumar, A., Rajasekar, B., Natarajan, R., Atajanova, A., & Makhmudov, S. (2025). Smart logistics integrating AI, IoT, blockchain, and Robotics for a sustainable Future. In *Emerging Trends in Smart Logistics Technologies* (pp. 73–104). IGI Global Scientific Publishing., DOI: 10.4018/979-8-3373-2434-0.ch003

Marzi, S., Mysiak, J., Essenfelder, A. H., Amadio, M., Giove, S., & Fekete, A. (2019). Constructing a comprehensive disaster resilience index: The case of Italy. *PLoS One, 14*(9), e0221585. DOI: 10.1371/journal.pone.0221585 PMID: 31525230

Masoum, A., Meratnia, N., & Havinga, P. J. M. (2018). Coalition formation based compressive sensing in wireless sensor networks. *Sensors (Basel), 18*(7), 2331. DOI: 10.3390/s18072331 PMID: 30021980

Mazoyer, M., & Roudart, L. (2006). *A History of World Agriculture: From the Neolithic Age to the Current Crisis*. Monthly Review Press.

Mehta, Y., Xu, R., Lim, B., Wu, J., & Gao, J. (2023). A review for green energy machine learning and AI services. *Energies, 16*(15), 5718. DOI: 10.3390/en16155718

Mengelkamp, E., Gärttner, J., Rock, K., Kessler, S., Orsini, L., & Weinhardt, C. (2018). Designing microgrid energy markets. *Applied Energy, 210*, 870–880. DOI: 10.1016/j.apenergy.2017.06.054

Microsoft & ICRISAT. (2018). *AI for Earth: Intelligent Tools for Sustainable Agriculture in India*. Microsoft Research India.

Molchanov, P., Tyree, S., Karras, T., Aila, T., & Kautz, J. (2019). Pruning convolutional neural networks for resource-efficient inference. *International Journal of Computer Vision, 128*(2), 501–520. DOI: 10.1007/s11263-019-01157-0

Mousavi, S. (2024). *AI game-changers for environmental accounting & sustainable finance*. Stanford Sustainable Finance Initiative.

Mukherji, A. (2023). Climate change 2023 synthesis report.

Nakamoto, S. (2008). Bitcoin: A peer-to-peer electronic cash system. https://bitcoin.org/bitcoin.pdf

Nasimian, A., Al Ashiri, L., Ahmed, M., Duan, H., Zhang, X., Rönnstrand, L., & Kazi, J. U. (2023). A receptor tyrosine kinase inhibitor sensitivity prediction model identifies AXL dependency in leukemia. *International Journal of Molecular Sciences*, *24*(4), 3830. DOI: 10.3390/ijms24043830 PMID: 36835239

National Oceanic and Atmospheric Administration. (n.d.). Numerical weather prediction (NWP). U.S. Department of Commerce. https://www.ncei.noaa.gov/products/weather-climate-models/numerical-weather- prediction

Nieto-Mora, D. A., Rodríguez-Buritica, S., Rodríguez-Marín, P., Martínez-Vargaz, J. D., & Isaza-Narváez, C. (2023). Systematic review of machine learning methods applied to ecoacoustics and soundscape monitoring. *Heliyon*, *9*(10), e20275. DOI: 10.1016/j.heliyon.2023.e20275 PMID: 37790981

Nkengla-Asi, L., Bernardini, M. D. R. C., Cohen, M. J., Lawson-Lartego, L., & Coates, K. (2024). Sustainable Development Goal 11: Make cities inclusive, safe, resilient, and sustainable. In *Handbook on Public Policy and Food Security* (pp. 268-280). Edward Elgar Publishing.

OECD. (2021). *Artificial intelligence, data governance and privacy: Policy considerations*. OECD Publishing.

Olivetti, L., & Messori, G. (2024). Advances and prospects of deep learning for medium-range extreme weather forecasting. *Geoscientific Model Development*, *17*(6), 2347–2358. DOI: 10.5194/gmd-17-2347-2024

Ouni, R., & Saleem, K. (2022). Framework for sustainable wireless sensor network based environmental monitoring. *Sustainability (Basel)*, *14*(14), 8356. DOI: 10.3390/su14148356

Pandurangan, A. P., Ochoa-Montano, B., Ascher, D. B., & Blundell, T. L. (2017). SDM: A server for predicting effects of mutations on protein stability. *Nucleic Acids Research*, *45*(W1), W229–W235. DOI: 10.1093/nar/gkx439 PMID: 28525590

Pathak, J., Subramanian, S., Harrington, P., Raja, S., Chattopadhyay, A., Mardani, M., Kurth, T., Hall, D., Li, Z., & Anandkumar, A. (2022). *FourCastNet: A global data-driven high-resolution weather model using adaptive Fourier neural operators*. arXiv. https://arxiv.org/abs/2202.11214

Pettorelli, N., & Williams, J. Schulte to Bühne, H., & Crowson, M. (2025). Deep learning and satellite remote sensing for biodiversity monitoring and conservation. *Remote Sensing in Ecology and Conservation, 11*(2), 123–13210.1002/rse2.415

Pingali, P. (2007). Agricultural Mechanization: Adoption Patterns and Economic Impact. Handbook of Agricultural Economics, Elsevier.

Poore, J., & Nemecek, T. (2018). Reducing food's environmental impacts through producers and consumers. *Science, 360*(6392), 987–992. DOI: 10.1126/science.aaq0216 PMID: 29853680

Pottier, C., Fresnais, M., Gilon, M., Jérusalem, G., Longuespée, R., & Sounni, N. E. (2020). Tyrosine kinase inhibitors in cancer: Breakthrough and challenges of targeted therapy. *Cancers (Basel), 12*(3), 731. DOI: 10.3390/cancers12030731 PMID: 32244867

Raman, R., Pattnaik, D., Lathabai, H. H., Kumar, C., Govindan, K., & Nedungadi, P. (2024). Green and sustainable AI research: An integrated thematic and topic modeling analysis. *Journal of Big Data, 11*(1), 55. DOI: 10.1186/s40537-024-00920-x

Rani, P., Dutta, K., & Kumar, V. (2022). Artificial intelligence techniques for prediction of drug synergy in malignant diseases: Past, present, and future. *Computers in Biology and Medicine, 144*, 105334. DOI: 10.1016/j.compbiomed.2022.105334 PMID: 35286893

Rasp, S., Dueben, P. D., Scher, S., Weyn, J. A., Mouatadid, S., & Thuerey, N. (2020). WeatherBench: A benchmark dataset for data-driven weather forecasting. *Journal of Advances in Modeling Earth Systems, 12*(11), e2020MS002203. DOI: 10.1029/2020MS002203

Rasp, S., Dueben, P. D., Scher, S., Weyn, J. A., Mouatadid, S., & Thuerey, N. (2023). WeatherBench 2: A benchmark for the next generation of data-driven weather models. *Journal of Advances in Modeling Earth Systems, 15*(2), e2022MS003156. DOI: 10.1029/2022MS003156

Ravikumar, A. S., Kalandarov, A., Djumaniyazova, M., & Habib, M. (2025). A roadmap to a sustainable future. In Advances in Computational Intelligence and Robotics (pp. 267–302). IGI Global Scientific Publishing. DOI: 10.4018/979-8-3373-3017-4.ch010

Ravikumar, R. N., & Aarthi, S. (2025a). AI-driven governance. In *Advances in Computational Intelligence and Robotics* (pp. 343–372). IGI Global Scientific Publishing., DOI: 10.4018/979-8-3373-3760-9.ch012

Ravikumar, R. N., & Aarthi, S. (2025b). Digital Twin-driven green energy systems. In *The Future of Green Energy* (pp. 1–28). IGI Global Scientific Publishing., DOI: 10.4018/979-8-3373-3226-0.ch001

RCMRD. (2019). *Africa Regional Data Cube: Earth Observation and Predictive Analytics for Drought Monitoring*. Regional Centre for Mapping of Resources for Development.

Reiche, D. (2010). Renewable energy policies in the Gulf countries: A case study of the carbon-neutral "Masdar City" in Abu Dhabi. *Energy Policy*, *38*(1), 378–382. DOI: 10.1016/j.enpol.2009.09.028

Reichstein, M., Camps-Valls, G., Stevens, B., Jung, M., Denzler, J., Carvalhais, N., & Prabhat, . (2019). Deep learning and process understanding for data-driven Earth system science. *Nature*, *566*(7743), 195–204. DOI: 10.1038/s41586-019-0912-1 PMID: 30760912

Rejeb, A., Keogh, J. G., & Treiblmaier, H. (2020). Leveraging the Internet of Things and Blockchain Technology in Supply Chain Management. *Future Generation Computer Systems*, *107*, 501–514.

Ren, X., Zhang, Y., & Liu, S. (2022). Energy-aware scheduling for AI model execution in renewable-powered systems. *IEEE Transactions on Sustainable Computing*, *7*(4), 842–853. DOI: 10.1109/TSUSC.2022.3149529

Roadmap, I. N. Z. (2023). *A Global Pathway to keep the 1.5° C Goal in Reach*. IEA.

Rolnick, D., Donti, P. L., Kaack, L. H., Kochanski, K., Lacoste, A., Sankaran, K., Ross, A. S., Milojevic-Dupont, N., Jaques, N., Waldman-Brown, A., Luccioni, A. S., Maharaj, T., Sherwin, E. D., Kording, K. P., Gomes, C., Ng, A. Y., & Bengio, Y. (2022). Tackling climate change with machine learning. *Nature Climate Change*, *12*(1), 22–27. DOI: 10.1038/s41558-021-01265-3

Rolnick, D., Donti, P. L., Kaack, L. H., Kochanski, K., Lacoste, A., Sankaran, K., Ross, A. S., Milojevic-Dupont, N., Jaques, N., Waldman-Brown, A., Luccioni, A. S., Maharaj, T., Sherwin, E. D., Mukkavilli, S. K., Kording, K. P., Gomes, C. P., Ng, A. Y., Hassabis, D., Platt, J. C., & Bengio, Y. (2023). Tackling climate change with machine learning. *ACM Computing Surveys*, *55*(2), 1–96. DOI: 10.1145/3485128

Rolnick, D., Donti, P. L., Kaack, L. H., Kochanski, K., Lacoste, A., Sankaran, K., Ross, A. S., Milojevic-Dupont, N., Jaques, N., Waldman-Brown, A., Luccioni, A., Maharaj, T., Sherwin, E. D., Mukkavilli, S. K., Kording, K. P., Gomes, C. P., Ng, A. Y., Hassabis, D., Platt, J. C., & Bengio, Y. (2022). Tackling climate change with machine learning. ACM Computing Surveys, 55(2). *Article. 42*, 1–96.

Ronco, M., Tárraga, J. M., Muñoz, J., Piles, M., Marco, E. S., Wang, Q., Espinosa, M. T. M., Ponserre, S., & Camps-Valls, G. (2023). Exploring interactions between socioeconomic context and natural hazards on human population displacement. *Nature Communications*, *14*(1), 8004. Advance online publication. DOI: 10.1038/s41467-023-43809-8 PMID: 38049446

Rosenzweig, S. A. (2018). Acquired resistance to drugs targeting tyrosine kinases. *Advances in Cancer Research*, *138*, 71–98. DOI: 10.1016/bs.acr.2018.02.003 PMID: 29551130

Roskoski, R.Jr. (2019). Properties of FDA-approved small molecule protein kinase inhibitors. *Pharmacological Research*, *144*, 19–50. DOI: 10.1016/j.phrs.2019.03.006 PMID: 30877063

Roskoski, R.Jr. (2021). Properties of FDA-approved small molecule protein kinase inhibitors: A 2021 update. *Pharmacological Research*, *165*, 105463. Advance online publication. DOI: 10.1016/j.phrs.2021.105463 PMID: 33513356

Rotz, S., Gravely, E., Mosby, I., Duncan, E., & Davison, J. (2019). The politics of digital agricultural platforms: Data access and power. *Geoforum*, *99*, 89–97.

Rutten, C. J., Velthuis, A. G. J., Steeneveld, W., & Hogeveen, H. (2013). Invited review: Sensors to support health management on dairy farms. *Journal of Dairy Science*, *96*(4), 1928–1952. DOI: 10.3168/jds.2012-6107 PMID: 23462176

Saberi, S., Kouhizadeh, M., Sarkis, J., & Shen, L. (2019). Blockchain technology and its relationships to sustainable supply chain management. *International Journal of Production Research*, *57*(7), 2117–2135. DOI: 10.1080/00207543.2018.1533261

Saptakee, S. (2024, December 9). *Green AI explained: Fueling innovation with a smaller carbon footprint*. CarbonCredits.com.

Schwartz, R., Dodge, J., Smith, N. A., & Etzioni, O. (2020). Green AI. *Communications of the ACM*, *63*(12), 54–63. DOI: 10.1145/3381831

Sgantzos, K., Al Hemairy, M. A., Tzavaras, P., & Stelios, S. (2023). Triple-Entry Accounting as a Means of Auditing Large Language Models. *Journal of Risk and Financial Management*, *16*(9), 383. DOI: 10.3390/jrfm16090383

Sgantzos, K., Grigg, I., & Al Hemairy, M. (2022). Multiple Neighborhood Cellular Automata as a Mechanism for Creating an AGI on a Blockchain. *Journal of Risk and Financial Management*, *15*(8), 360. DOI: 10.3390/jrfm15080360

Shah, M. C. (2019). Multi-objective optimization of sustainable building designs for energy consumption using Ai -ml framework. *International Journal of Psychosocial Rehabilitation*, •••, 1224–1234. DOI: 10.53555/V23I3/400142

Shamshiri, R. R.. (2018). Robotics and automation for sustainable agriculture. *Engineering in Agriculture, Environment and Food*, *11*(2), 64–75.

Shamshiri, R. R., Kalantari, F., Ting, K. C., Thorp, K. R., & Hameed, I. A. (2018). Advances in greenhouse automation and controlled-environment agriculture: A review. *Computers and Electronics in Agriculture*, *162*, 262–277.

Shankar, R., & Xu, Q. (2025). Automated detection of greenwashing in Indian corporate sustainability reports using natural language processing. DOI: 10.2139/ssrn.5294480

Sharjah Fish Resources Authority. (2025). Sharjah Fish Resources Authority. Government of Sharjah. https://fra.shj.ae/en

Shi, W., Cao, J., Zhang, Q., Li, Y., & Xu, L. (2019). Edge computing: Vision and challenges. *IEEE Internet of Things Journal*, *3*(5), 637–646. DOI: 10.1109/JIOT.2016.2579198

Shiwlani, A., Kumar, S., Hasan, S. U., Kumar, S., & Naguib, J. S. (2024). Advancing Hepatology with AI: A Systematic Review of Early Detection Models for Hepatitis-Associated Liver Cancer. [IJISRT]. *International Journal of Innovative Science and Research Technology*, *9*(11), 2522–2536.

Singha, M., Pu, L., Stanfield, B. A., Uche, I. K., Rider, P. J., Kousoulas, K. G., Ramanujam, J., & Brylinski, M. (2022). Artificial intelligence to guide precision anticancer therapy with multitargeted kinase inhibitors. *BMC Cancer*, *22*(1), 1211. DOI: 10.1186/s12885-022-10293-0 PMID: 36434556

Singh, V., & Agnihotri, A. (2024). Addressing environmental challenges through artificial intelligence (AI)-powered natural disaster management. *International Journal of Applied and Scientific Research*, *2*(5), 485–496. DOI: 10.59890/ijasr.v2i5.1413

Snider, L. (2025). *NSF NCAR is making AI weather modeling easier to access and use*.

Solaw, F. A. O. (2011). *The state of the world's land and water resources for food and agriculture*.

Spencer, G. (2018). *AI for Earth: Helping save the planet with data science*.

StandfordUniversity. (2023). Artificial Intelligence Index Report Introduction to the AI Index Report 2023 GP-003. *Human-Centered Artificial Intelligence*, 1–386.

Stern, N., & Romani, M. (2025, January). What is AI's role in the climate transition and how can it drive growth? In Opinion, World Economic Forum, January (Vol. 16).

Strubell, E., Ganesh, A., & McCallum, A. (2020, April). Energy and policy considerations for modern deep learning research. In *Proceedings of the AAAI conference on artificial intelligence* (Vol. 34, No. 09, pp. 13693 13696). DOI: 10.1609/aaai.v34i09.7123

Strubell, E., Ganesh, A., & McCallum, A. (2019). Energy and policy considerations for deep learning in NLP. *Proceedings of ACL*, 3645-3650. DOI: 10.18653/v1/P19-1355

Sun, X., & Hu, B. (2018). Mathematical modeling and computational prediction of cancer drug resistance. *Briefings in Bioinformatics*, *19*(6), 1382–1399. DOI: 10.1093/bib/bbx065 PMID: 28981626

Tabbakh, A., Al Amin, L., Islam, M., Mahmud, G. M., Chowdhury, I. K., & Mukta, M. S. (2024). Towards sustainable AI: A comprehensive framework for green AI. *Discover Sustainability*, *5*(1), 408. Advance online publication. DOI: 10.1007/s43621-024-00641-4

Tang, Y., Sun, Y., Han, Z., Soomro, S., Wu, Q., Tan, B., & Hu, C. (2023). flood forecasting based on machine learning pattern recognition and dynamic migration of parameters. *Journal of Hydrology. Regional Studies*, *47*, 101406. DOI: 10.1016/j.ejrh.2023.101406

Task Force on Climate-related Financial Disclosures. (2023). *2023 status report*. Financial Stability Board.

Thulke, D., Gao, Y., Pelser, P., Brune, R., Jalota, R., Fok, F., Ramos, M., van Wyk, I., Nasir, A., Goldstein, H., Tragemann, T., Nguyen, K., Fowler, A., Stanco, A., Gabriel, J., Taylor, J., Moro, D., Tsymbalov, E., de Waal, J., … Erasmus, D. (2024). *ClimateGPT: Towards AI Synthesizing Interdisciplinary Research on Climate Change*. 1–52.

Tilman, D., Cassman, K. G., Matson, P. A., Naylor, R., & Polasky, S. (2002). Agricultural sustainability and intensive production practices. *Nature*, *418*(6898), 671–677. DOI: 10.1038/nature01014 PMID: 12167873

Tscharntke, T., Clough, Y., Wanger, T. C., Jackson, L., Motzke, I., Perfecto, I., Vandermeer, J., & Whitbread, A. (2012). Global food security, biodiversity conservation and the future of agricultural intensification. *Biological Conservation*, *151*(1), 53–59. DOI: 10.1016/j.biocon.2012.01.068

Turner, W., Rondinini, C., Pettorelli, N., Mora, B., & Leidner, A. K. (2015). Remote sensing for biodiversity science and conservation. *Trends in Ecology & Evolution, 30*(7), 403–414.

UNDRR. (2019). *Global Assessment Report on Disaster Risk Reduction*. United Nations Office for Disaster Risk Reduction.

United Nations Office for Disaster Risk Reduction. (2023). *Global Assessment Report on Disaster Risk Reduction 2023: Mapping Resilience for the Sustainable Development Goals*. Stylus Publishing, LLC.

Vaghefi, S. A., Stammbach, D., Muccione, V., Bingler, J., Ni, J., Kraus, M., Allen, S., Colesanti-Senni, C., Wekhof, T., Schimanski, T., Gostlow, G., Yu, T., Wang, Q., Webersinke, N., Huggel, C., & Leippold, M. (2023). ChatClimate: Grounding conversational AI in climate science. *Communications Earth & Environment, 4*(1), 480. Advance online publication. DOI: 10.1038/s43247-023-01084-x

Verdecchia, R., Sallou, J., & Cruz, L. (2023). A systematic review of Green AI. *Wiley Interdisciplinary Reviews. Data Mining and Knowledge Discovery, 13*(4), e1507. Advance online publication. DOI: 10.1002/widm.1507

Vinuesa, R., Azizpour, H., Leite, I., Balaam, M., Dignum, V., Domisch, S., & Nerini, F. F. (2020). The role of artificial intelligence in achieving the Sustainable Development Goals. *Nature Communications, 11*(1), 233. DOI: 10.1038/s41467-019-14108-y PMID: 31932590

Von Struensee, S. (2021). Artificial intelligence and corporate social responsibility: Employees' key role in driving responsible artificial intelligence at big tech. SSRN *Electronic Journal*. DOI: 10.2139/ssrn.3873097

Wäldchen, J., & Mäder, P. (2018). Machine learning for biodiversity research. *Ecological Informatics, 45*, 88–95.

Wäldchen, J., & Mäder, P. (2018). Machine learning for image-based species identification. *Ecological Informatics, 45*, 88–95.

Wamba, S. F., Gunasekaran, A., Akter, S., Ren, S. J., Dubey, R., & Childe, S. J. (2020). Big data analytics and firm performance: Effects of dynamic capabilities. *Journal of Business Research, 111*, 88–100.

Wang, Z., Jang, W., Ruan, B., Wang, J., & Xiao, S. (2025). Developing and Integrating Trust Modeling into Multi Objective Reinforcement Learning for Intelligent Agricultural Management. *arXiv preprint arXiv:2505.10803*.

Wang, C., & Zhang, Y. (2017). Improving scoring-docking-screening powers of protein–ligand scoring functions using random forest. *Journal of Computational Chemistry, 38*(3), 169–177. DOI: 10.1002/jcc.24667 PMID: 27859414

Wang, N., Pan, H., Feng, Y., & Du, S. (2023). How do ESG practices create value for businesses? Research review and prospects. *Sustainability Accounting. Management and Policy Journal, 15*(5), 1155–1177. DOI: 10.1108/SAMPJ-12-2021-0515

Wang, Q., Qi, Y., & Li, R. (2025). Artificial intelligence and corporate sustainability: Shaping the future of <scp>ESG</scp> in the age of industry 5.0. *Sustainable Development.* Advance online publication. DOI: 10.1002/sd.70200

Wang, S., Zheng, Z., Li, X., & Chen, X. (2023). Knowledge distillation for efficient weather forecasting with deep learning models. *Environmental Modelling & Software, 166*, 105683. DOI: 10.1016/j.envsoft.2023.105683

Wibowo, A., Amri, I., Surahmat, A., & Rusdah, R. (2025). Leveraging artificial intelligence in disaster management: A comprehensive bibliometric review. *Jàambá, 17*(1), 1776. DOI: 10.4102/jamba.v17i1.1776 PMID: 40357012

Wijaya, B., Rahayu, I. P., & Wijayanto, H. (2025). Integration of Federated Learning in big data analytics for IoT-based intelligent transportation system. *Infact: International Journal of Computers, 9*(01), 19–24. DOI: 10.61179/infact.v9i01.704

Wolfert, S., Ge, L., Verdouw, C., & Bogaardt, M. J. (2017). Big Data in Smart Farming – A Review. *Agricultural Systems, 153*, 69–80. DOI: 10.1016/j.agsy.2017.01.023

World Bank. (2018). *Africa Regional Data Cube (ARDC): Transforming Data for Sustainable Development.* World Bank Group.

World Bank. (2020). *Future of Food: Harnessing Digital Technologies to Improve Food System Outcomes.* World Bank Group.

World Economic Forum. (2023a). Blockchain for climate action. World Economic Forum.

World Economic Forum. (2023b). Harnessing artificial intelligence for climate and sustainability. World Economic Forum.

Wu, Y., & Xue, W. (2024). Data-Driven Weather Forecasting and Climate Modeling from the Perspective of Development. In *Atmosphere* (Vol. 15, Issue 6). DOI: 10.3390/atmos15060689

Wu, T., Wang, Y., Cui, X., Xue, P., & Qiao, Y. (2025). AI-based identification method for cervical transformation zone within digital colposcopy: Development and multicenter validation study. *JMIR Cancer, 11*, e69672. DOI: 10.2196/69672 PMID: 40163848

Wu, Z., Yin, H., He, H., & Li, Y. (2022). Dynamic-LSTM hybrid models to improve seasonal drought predictions over China. *Journal of Hydrology. Regional Studies, 615*(128706).

Xu, G., Lu, Y., Zhang, L., & Vasilakos, A. (2021). Edge Intelligence: Architectures, Challenges, and Applications. *IEEE Internet of Things Journal, 8*(10), 7151–7172.

Xu, H., Chen, J., Wang, Y., Huang, X., & Wang, Z. (2023). Low-rank adaptation for efficient deep learning. *IEEE Transactions on Neural Networks and Learning Systems, 34*(9), 6062–6074. DOI: 10.1109/TNNLS.2023.3244421 PMID: 34932488

Xu, X., Weber, I., & Staples, M. (2019). *Architecture for blockchain applications*. Springer. DOI: 10.1007/978-3-030-03035-3

Xu, Y., Liu, H., & Zhao, M. (2021). Lightweight deep learning models for energy-efficient weather forecasting. *Environmental Informatics Letters, 5*(2), 80–88. DOI: 10.1016/j.eil.2021.06.002

Xu, Y., Liu, X., Cao, X., Huang, C., Liu, E., Qian, S., Liu, X., Wu, Y., Dong, F., Qiu, C. W., Qiu, J., Hua, K., Su, W., Wu, J., Xu, H., Han, Y., Fu, C., Yin, Z., Liu, M., & Zhang, J. (2021). Artificial intelligence: A powerful paradigm for scientific research. *Innovation (Cambridge (Mass.)), 2*(4), 100179. DOI: 10.1016/j.xinn.2021.100179 PMID: 34877560

Yanfei, J., Guangda, C., & Liquan, Z. (2021). Energy-efficient routing protocol based on zone for heterogeneous wireless sensor networks. *Journal of Electrical and Computer Engineering, 2021*, 1–9. DOI: 10.1155/2021/5557756

Yang, H., Li, D., Zhao, X., & Chen, Y. (2025). Artificial intelligence-driven aquaculture: Applications, challenges, and sustainability implications. *Aquaculture Reports, 27*, 101373.

Yang, Z. Y., Ye, Z. F., Xiao, Y. J., Hsieh, C. Y., & Zhang, S. Y. (2022). SPLDExtraTrees: Robust machine learning approach for predicting kinase inhibitor resistance. *Briefings in Bioinformatics, 23*(3), bbac050. Advance online publication. DOI: 10.1093/bib/bbac050 PMID: 35262669

Yonetani, R., Lim, S. N., & Ahn, J. (2021). Distributed and federated learning for real-time environmental prediction. *ACM Transactions on Sensor Networks, 17*(4), 1–24. DOI: 10.1145/3470913

Zaman, S. I., Jadoon, S. T., & Khan, S. A. (2024). Green algorithms: The impact of artificial intelligence on environmental sustainability. *1st International Conference on Industrial, Manufacturing, and Process Engineering (ICIMP-2024)*, 40. DOI: 10.3390/engproc2024076040

Zarei, P., & Ghasemi, F. (2024). The application of artificial intelligence and drug repositioning for the identification of fibroblast growth factor receptor inhibitors: A review. *Advanced Biomedical Research*, *13*, 9. DOI: 10.4103/abr.abr_170_23 PMID: 38525398

Zhang, H., Liu, Y., & Feng, Z. (2023). Energy-efficient edge-cloud collaboration for intelligent weather forecasting. *Future Generation Computer Systems*, *145*, 147–160. DOI: 10.1016/j.future.2023.03.007

Zhang, Y.. (2022). Blockchain-based carbon trading mechanisms and policy implications. *Energy Policy*, *162*, 112776.

Zhou, Y., Portelli, S., Pat, M., Rodrigues, C. H., Nguyen, T.-B., Pires, D. E., & Ascher, D. B. (2021). Structure-guided machine learning prediction of drug resistance mutations in Abelson 1 kinase. *Computational and Structural Biotechnology Journal*, *19*, 5381–5391. DOI: 10.1016/j.csbj.2021.09.016 PMID: 34667533

Zhu, H., & Jin, Y. (2020). Multi-objective evolutionary federated learning. *IEEE Transactions on Neural Networks and Learning Systems*, *31*(4), 1310–1322. DOI: 10.1109/TNNLS.2019.2919699 PMID: 31247576

Zou, W., Li, L., Xu, Z., Wu, D., Xu, C., Wang, Y., Zhu, H., & Sun, X. (2023). SmartDL: Energy-aware decremental learning in a mobile-based federation for geo-spatial system. *Neural Computing & Applications*, *35*(5), 3677–3696. DOI: 10.1007/s00521-021-06378-9

About the Contributors

Fanar Shwedeh is a Computer Science specialist and a researcher working in the area of AI Innovations, Information Systems and Technology Management besides teaching as an Associate Professor in City University Ajman. Holds a PhD degree in Technology Management Information Systems from Utara University Malaysia, Masters in Business Management, BSc in Computer Science from the University of Sharjah and high Diploma in Cyber Security of Information Systems from the USA Cyber Security Institution. Dr. Fanar has published research publications in highly impacted international journals, and she is an academic chief reviewer of reputable International Scopus-indexed journals. Dr. Fanar is a member of the research academic unit in Dubai Smart government office. She was awarded the H.H. Sheikh Hamdan Bin Rashid Al-Maktoum Award for Distinguished Performance in 2020. Furthermore, she is the editor of the book chapter "Sustainable Leadership for Environmental Risk", Springer 2025.

***

Alaina Thea V. Concepcion is an Associate Professor at Bulacan State University, Philippines. She has been teaching for over fourteen years, handling undergraduate students in the field of Information Technology. Her academic experience includes instruction in computing, systems development, and emerging technologies, with a strong commitment to student learning and academic excellence. As an educator, she is dedicated to continuously improving her teaching practice and contributing to research and academic initiatives that support the advancement of information technology education.

Anthony Concepcion is an Associate Professor at Bulacan State University, Philippines. He has been teaching Information Technology courses at the collegiate level for over eighteen years. He has also served as Program Chair of the Information

Technology Department, where he contributed to curriculum development, academic leadership, and program implementation. His professional work reflects a strong commitment to quality instruction, academic service, and the continuous improvement of information technology education.

Umamaheswari Purushothaman holds a B.E. in Information Technology from Periyar Maniammai College of Engineering and Technology (2003), an M.B.A. from Alagappa University, Karaikudi, Tamil Nadu (2006), and an M.Tech in Computer Science from SASTRA Deemed University, Thanjavur, Tamil Nadu (2011). She earned her Ph.D. in Computer Science and Engineering from SASTRA Deemed University. With two decades of experience, she has taught both undergraduate and postgraduate courses in Computer Science. She is currently an Assistant Professor in the Department of Computer Science and Engineering at SRC Campus, SASTRA University, Kumbakonam. Her research interests focus on Data Mining, Machine Learning, and Deep Learning. She has published numerous research papers in international conferences, SCI-indexed journals, Scopus, and book chapters.

Ravikumar R. N. is currently a Postdoctoral researcher at Universiti Malaysia Perlis, Malaysia. Also, working as an Associate Professor in the department of Computer Engineering at Marwadi University, Rajkot, Gujarat. He has 12+ years of national and international teaching experience. He completed his PhD at Amity University, Rajasthan, M.E at VMRF Deemed University, B.E at Anna University, Chennai. He is the co-founder of the Research Activities Club at Marwadi University and trains students on research activities, which leads to the publication of 7 Scopus/SCI journals, 20 Book chapters, 25 Conference papers, and 10 patents. His specialization is ML, DL, Federated Learning, Recommendation Systems, and Optimization Techniques.

Aarthi S is a dedicated Assistant Professor in Computer Science with over six years of teaching experience. She is currently pursuing her Ph.D. at Marwadi University, Gujarat, and holds an M.Phil. and MCA from Bharathiar University, Tamil Nadu. Throughout her academic career, Aarthi has specialized in programming languages such as C++, Java, and C, and contributed to research in recommender systems and AI-related data privacy. She has published journal articles, contributed to book chapters, and actively participated in technical workshops. Known for her student mentorship and engagement in academic events, Aarthi remains committed to fostering innovation and excellence in computer science education.

Binastya Anggara Sekti has been a lecturer at the Faculty of Computer Science, Esa Unggul University, Jakarta, since 2016. He is actively involved in the Artificial Intelligence and Information Security Study Group. He holds a Bachelor's degree in

Informatics Engineering from the Bandung Institute of Technology and a Master's degree in Information Systems Management from Trisakti University, Jakarta. Prior to entering academia, he accumulated over 25 years of professional experience as an IT practitioner in various companies across Jakarta. His work focused on developing data centers for internet service providers and delivering IT solutions for the manufacturing and oil & gas sectors. His research interests include artificial intelligence, cybersecurity, and the development of IT infrastructure.

Index

A

B

C

D

E

F

www.ingramcontent.com/pod-product-compliance
Lightning Source LLC
LaVergne TN
LVHW081523300926
849161LV00006B/1278